Westfield Memorial Library
Westfield, New Jers

S0-BCU-550

KLUGE

BOOKS BY GARY MARCUS

The Algebraic Mind

The Birth of the Mind

The Norton Psychology Reader

Kluge

Westfield Memorial Library
Westfield, New Jersey

KLUGE

THE HAPHAZARD CONSTRUCTION
OF THE HUMAN MIND

GARY MARCUS

HOUGHTON MIFFLIN COMPANY

BOSTON · NEW YORK   2008

Copyright © 2008 by Gary Marcus

All rights reserved

For information about permission to reproduce selections from
this book, write to Permissions, Houghton Mifflin Company,
215 Park Avenue South, New York, New York 10003.

www.houghtonmifflinbooks.com

Library of Congress Cataloging-in-Publication Data

Marcus, Gary F. (Gary Fred)

Kluge : the haphazard construction of the human
mind / Gary Marcus.

  p. cm.

Includes bibliographical references (p.   ) and index.

ISBN 978-0-618-87964-9

1. Psychology. 2. Cognitive psychology. 3. Cognitive
neuroscience. I. Title.

BF38.M355 2008 153 — dc22

Printed in the United States of America

MP 10 9 8 7 6 5 4 3 2 1

Lines from "Laments for a Dying Language" copyright © 1960
by Ogden Nash. Reprinted by permission of Curtis Brown Ltd.

For my father,
who taught me the word

Contents

Living organisms are historical structures: literally creations of history. They represent not a perfect product of engineering, but a patchwork of odd sets pieced together when and where opportunities arose.

— FRANÇOIS JACOB

Bad luck is better than no luck at all.

— TRADITIONAL

KLUGE

1

REMNANTS OF HISTORY

It has been said that man is a rational animal. All my life I have
been searching for evidence which could support this.

— BERTRAND RUSSELL

ARE HUMAN BEINGS "noble in reason" and "infinite in faculty" as
William Shakespeare famously wrote? Perfect, "in God's image," as
some biblical scholars have asserted? Hardly.

If mankind were the product of some intelligent, compassionate
designer, our thoughts would be rational, our logic impeccable. Our
memory would be robust, our recollections reliable. Our sentences
would be crisp, our words precise, our languages systematic and
regular, not besodden with irregular verbs (*sing-sang, ring-rang,* yet
bring-brought) and other peculiar inconsistencies. As the language
maven Richard Lederer has noted, there would be ham in hamburger,
egg in eggplant. English speakers would park in parkways and drive
on driveways, and not the other way around.

At the same time, we humans are the only species smart enough
to systematically plan for the future — yet dumb enough to ditch our
most carefully made plans in favor of short-term gratification. ("Did
I say I was on a diet? *Mmm,* but three-layer chocolate mousse is my
favorite . . . Maybe I'll start my diet *tomorrow.*") We happily drive
across town to save $25 on a $100 microwave but refuse to drive the
same distance to save exactly the same $25 on a $1,000 flat-screen TV.
We can barely tell the difference between a valid syllogism, such as *All
men are mortal, Socrates is a man, therefore Socrates is mortal,* and a
fallacious counterpart, such as *All living things need water, roses need*

water, therefore roses are living things (which seems fine until you substitute *car batteries* for *roses*). If I tell you that "Every sailor loves a girl," you have no idea whether I mean one girl in particular (say, Betty Sue) or whether I'm really saying "to each his own." And don't even get me started on eyewitness testimony, which is based on the absurd premise that we humans can accurately remember the details of a briefly witnessed accident or crime, years after the fact, when the average person is hard pressed to keep a list of a dozen words straight for half an hour.

I don't mean to suggest that the "design" of the human mind is a total train wreck, but if I were a politician, I'm pretty sure the way I'd put it is "mistakes were made." The goal of this book is to explain what mistakes were made — and why.

Where Shakespeare imagined infinite reason, I see something else, what engineers call a "kluge." A kluge is a clumsy or inelegant — yet surprisingly effective — solution to a problem. Consider, for example, what happened in April 1970 when the CO_2 filters on the already endangered lunar module of *Apollo 13* began to fail. There was no way to send a replacement filter up to the crew — the space shuttle hadn't been invented yet — and no way to bring the capsule home for several more days. Without a filter, the crew would be doomed. The mission control engineer, Ed Smylie, advised his team of the situation, and said, in effect, "Here's what's available on the space capsule; figure something out." Fortunately, the ground crew was able to meet the challenge, quickly cobbling together a crude filter substitute out of a plastic bag, a cardboard box, some duct tape, and a sock. The lives of the three astronauts were saved. As one of them, Jim Lovell, later recalled, "The contraption wasn't very handsome, but it worked."

Not every kluge saves lives. Engineers sometimes devise them for sport, just to show that something — say, building a computer out of Tinkertoys — can be done, or simply because they're too lazy to do something the right way. Others cobble together kluges out of a mixture of desperation and resourcefulness, like the TV character MacGyver, who, needing to make a quick getaway, jerry-built a pair of shoes from duct tape and rubber mats. Other kluges are created just for laughs, like Wallace and Gromit's "launch and activate" alarm clock/coffeemaker/Murphy bed and Rube Goldberg's "simplified pencil sharpener" (a kite attached to a string lifts a door, which allows moths to escape, culminating in the lifting of a cage, which frees a woodpecker to gnaw the wood that surrounds a pencil's graphite core). MacGyver's shoes and Rube Goldberg's pencil sharpeners are nothing, though, compared to perhaps the most fantastic kluge of them all — the human mind, a quirky yet magnificent product of the entirely blind process of evolution.

The origin, and even the spelling, of the word *kluge* is up for grabs. Some spell it with a *d* (*kludge*), which has the virtue of looking as clumsy as the solutions it denotes, but the disadvantage of suggesting the wrong pronunciation. (Properly pronounced, *kluge* rhymes with *huge*, not *sludge*.)* Some trace the word to the old Scottish word *cludgie*, which means "an outside toilet." Most believe the origins lie in the German word *Kluge*, which means "clever." *The Hacker's Dictionary of Computer Jargon* traces the term back at least to 1935, to a "Kluge [brand] paper feeder," described as "an adjunct to mechanical printing presses."

> The Kluge feeder was designed before small, cheap electric motors and control electronics; it relied on a fiendishly complex assortment of cams, belts, and linkages to both power and synchronize all its operations from one motive driveshaft. It was accordingly temperamental, subject to frequent breakdowns, and devilishly difficult to repair — but oh, so clever!

*One could argue that the spelling *klooge* (rhymes with *stooge*) would even better capture the pronunciation, but I'm not about to foist a third spelling upon the world.

Virtually everybody agrees that the term was first popularized in February 1962, in an article titled "How to Design a Kludge," written, tongue in cheek, by a computer pioneer named Jackson Granholm, who defined a kluge as "an ill-assorted collection of poorly matching parts, forming a distressing whole." He went on to note that "the building of a Kludge . . . is not work for amateurs. There is a certain, indefinable, masochistic finesse that must go into true Kludge building. The professional can spot it instantly. The amateur may readily presume that 'that's the way computers are.'"

The engineering world is filled with kluges. Consider, for example, something known as vacuum-powered windshield wipers, common in most cars until the early 1960s. Modern windshield wipers, like most gizmos on cars, are driven by electricity, but back in the olden days, cars ran on 6 volts rather than 12, barely enough power to keep the spark plugs going and certainly not enough to power luxuries like windshield wipers. So some clever engineer rigged up a kluge that powered windshield-wiper motors with suction, drawn from the engine, rather than electricity. The only problem is that the amount of suction created by the engine varies, depending on how hard the engine is working. The harder it works, the less vacuum it produces. Which meant that when you drove your 1958 Buick Riviera up a hill, or accelerated hard, your wipers slowed to a crawl, or even stopped working altogether. On a rainy day in the mountains, Grandpa was out of luck.

What's really amazing — in hindsight — is that most people probably didn't even realize it was possible to do better. And this, I think, is a great metaphor for our everyday acceptance of the idiosyncrasies of the human mind. The mind is inarguably impressive, a lot better than any available alternative. But it's still flawed, often in ways we scarcely recognize. For the most part, we simply accept our faults — such as our emotional outbursts, our mediocre memories, and our vulnerability to prejudice — as standard equipment. Which is exactly why recognizing a kluge, and how it might be improved upon, sometimes requires thinking outside the box. The best science, like the best engineering, often comes from understanding not just how things are, but how else they could have been.

If engineers build kluges mostly to save money or to save time, why does nature build them? Evolution is neither clever nor penny-pinching. There's no money involved, no foresight, and if it takes a billion years, who's going to complain? Yet a careful look at biology reveals kluge after kluge. The human spine, for example, is a lousy solution to the problem of supporting the load in an upright, two-legged creature. It would have made a lot more sense to distribute our weight across four equal cross-braced columns. Instead, all our weight is borne by a single column, putting enormous stress on the spine. We manage to survive upright (freeing our hands), but the cost for many people is agonizing back pain. We are stuck with this barely adequate solution not because it is the best possible way to support the weight of a biped, but because the spine's structure evolved from that of four-legged creatures, and standing up poorly is (for creatures like us, who use tools) better than not standing up at all.

Meanwhile, the light-sensitive part of our eye (the retina) is installed backward, facing the back of the head rather than the front. As a result, all kinds of stuff gets in its way, including a bunch of wiring that passes through the eye and leaves us with a pair of blind spots, one in each eye.

Another well-known example of an evolutionary kluge comes from a rather intimate detail of male anatomy. The tubing that runs from the testis to the urethra (the vas deferens) is much longer than necessary: it runs back to front, loops around, and does a 180-degree turn back to the penis. A parsimonious designer interested in conserving materials (or in efficiency of delivery) would have connected the testis directly to the penis with just a short length of tubing; only because biology builds on what has come before is the system set up so haphazardly. In the words of one scientist, "The [human] body is a bundle of imperfections, with . . . useless protuberances above the nostrils, rotting teeth with trouble-prone third molars, aching feet . . . , easily strained backs, and naked tender skin, subject to cuts, bites, and, for many, sunburn. We are poor runners and are only about a third as strong as chimpanzees smaller than ourselves."

To this litany of human-specific imperfections, we might add dozens more that are widely shared across the animal world, such as the byzantine system by which DNA strands are separated prior to DNA replication (a key process in allowing one cell to become two). One molecule of DNA polymerase does its job in a perfectly straightforward fashion, but the other does so in a back-and-forth, herky-jerky way that would drive any rational engineer insane.

Nature is prone to making kluges because it doesn't "care" whether its products are perfect or elegant. If something works, it spreads. If it doesn't work, it dies out. Genes that lead to successful outcomes tend to propagate; genes that produce creatures that can't cut it tend to fade away; all else is metaphor. Adequacy, not beauty, is the name of the game.

Nobody would doubt this when it comes to the body, but somehow, when it comes to the mind, many people draw the line. Sure, my spine is a kluge, maybe my retina too, but my *mind*? It's one thing to accept that our body is flawed, quite another to accept that our mind is too.

Indeed, there is a long tradition in thinking otherwise. Aristotle saw man as "the rational animal," and economists going back to John Stuart Mill and Adam Smith have supposed that people make decisions based on their own self-interest, preferring wherever possible to buy low and sell high, maximizing their "utility" wherever they can.

In the past decade, a number of academics have started to argue that humans reason in a "Bayesian"* fashion, which is mathemati-

---

*The term *Bayesian* comes from a particular mathematical theorem stemming from the work of the Reverend Thomas Bayes (1702–1761), although he himself did not propose it as a model for human reasoning. In rough terms, the theorem states that the probability of some event is proportional to the product of the likelihood of that event and its prior probability. For a clear (though somewhat technical) introduction, point your browser to http://en.wikipedia.org/wiki/Bayesian_statistics.

cally optimal. One prestigious journal recently devoted an entire issue to this possibility, with a trio of prominent cognitive scientists from MIT, UCLA, and University College London arguing that "it seems increasingly plausible that human cognition may be explicable in rational probabilistic terms . . . in core domains, human cognition approaches an optimal level of performance."

The notion of optimality is also a recurrent theme in the increasingly popular field of evolutionary psychology. For example, John Tooby and Leda Cosmides, the cofounders of the field, have written that "because natural selection is a hill-climbing process that tends to choose the best of the variant designs that actually appear, and because of the immense numbers of alternatives that appear over the vast expanse of evolutionary time, natural selection tends to cause the accumulation of *superlatively well engineered* functional designs."

In the same vein, Steven Pinker has argued that "the parts of the mind that allow us to see are indeed well engineered, and there is no reason to think that the quality of engineering progressively deteriorates as the information flows upstream to the faculties that interpret and act on what we see."

This book will present a rather different view. Although no reasonable scholar would doubt the fact that natural selection *can* produce superlatively well engineered functional designs, it is also clear that superlative engineering is by no means *guaranteed*. What I will argue, in contrast to most economists, Bayesians, and evolutionary psychologists, is that the human mind is no less of a kluge than the body. And if that's true, our very understanding of ourselves — of human nature — must be reconsidered.

In the extensive literature on evolutionary psychology, I know of only a few aspects of the human mind that have been attributed to genuine quirks. Although most evolutionary psychologists recognize the possibility of suboptimal evolution *in principle*, in practice, when human errors are discussed, it's almost always to explain why something apparently nonadaptive actually turns out to be *well* engineered.

Take, for example, infanticide. Nobody would argue that infanticide is morally justifiable, but why does it happen at all? From the perspective of evolution, infanticide is not just immoral, but puzzling. If we exist essentially as gene-propagating vessels (as Richard Dawkins has argued), why would any parent murder his or her own child? Martin Daly and Margo Wilson have argued that from the gene's-eye view, infanticide makes sense only in a very limited set of circumstances: when the parent is not actually related by blood to the child (stepparents, for example), when a male parent is in doubt about paternity, or when a mother is not currently in a position to take good care of the child, yet has prospects for taking better care of some future child (say, because the current infant was born hopelessly unhealthy). As Daly and Wilson have shown, patterns of murder and child abuse fit well with these hypotheses.

Or consider the somewhat unsurprising fact that men (but not women) systematically tend to overinterpret the sexual intentions of potential mates.* Is this simply a matter of wishful thinking? Not at all, argue the evolutionary psychologists Martie Haselton and David Buss. Instead, it's a highly efficient strategy shaped by natural selection, a cognitive error reinforced by nature. Strategies that lead to greater reproductive success spread (by definition) widely throughout the population, and ancestral males who tended to read too much into the signals given by possible partners would have more opportunities to reproduce than would their more cautious counterparts, who likely failed to identify bona fide opportunities. From the gene's-eye view, it was well worth it for our male ancestors to take the risk of overinterpretation because gaining an extra reproductive opportunity far outweighs the downside, such as damage to self-esteem or reputation, of perceiving opportunity where there is none. What looks like a bug, a systematic bias in interpreting the motives of other human beings, might in this case actually be a positive feature.

When reading clever, carefully argued examples like this one, it's

---

*Except, tellingly, those of their sisters.

easy to get caught up in the excitement, to think that behind every human quirk or malfunction is a truly adaptive strategy. Underpinning such examples is a bold premise: that optimization is the inevitable outcome of evolution. But optimization is not an *inevitable* outcome of evolution, just a *possible* one. Some apparent bugs may turn out to be advantages, but — as the spine and inverted retina attest — some bugs may be genuinely suboptimal and remain in place because evolution just didn't find a better way.

Natural selection, the key mechanism of evolution, is only as good as the random mutations that arise. If a given mutation is beneficial, it may propagate, but the most beneficial mutations imaginable sometimes, alas, never appear. As an old saying puts it, "Chance proposes and nature disposes"; a mutation that does not arise cannot be selected for. If the right set of genes falls into place, natural selection will likely promote the spread of those genes, but if they don't happen to occur, all evolution can do is select the next best thing that's available.

To think about this, it helps to start with the idea of evolution as mountain climbing. Richard Dawkins, for example, has noted that there is little chance that evolution would assemble any complex creature or organ (say, the eye) overnight — too many lucky chance mutations would need to occur simultaneously. But it is possible to achieve perfection incrementally. In the vivid words of Dawkins,

> you don't need to be a mathematician or physicist to calculate that an eye or a hemoglobin molecule would take from here to infinity to self-assemble by sheer higgledy-piggledy luck. Far from being a difficulty peculiar to Darwinism, the astronomic improbability of eyes and knees, enzymes and elbow joints and the other living wonders is precisely the problem that any theory of life must solve, and that Darwinism uniquely does solve. It solves it by breaking the improbability up into small, manageable parts, smearing out the luck needed, going round the back of

Mount Improbable and crawling up the gentle slopes, inch by million-year inch.

And, to be sure, examples of sublime evolution abound. The human retina, for example, can detect a single photon in a darkened room, and the human cochlea (the hair cell containing the part of the inner ear that vibrates in response to sound waves) can, in an otherwise silent room, detect vibrations measuring less than the diameter of a hydrogen atom. Our visual systems continue, despite remarkable advances in computer power, to far outstrip the visual capacities of any machine. Spider silk is stronger than steel and more elastic than rubber. All else being equal, species (and the organs they depend upon) tend, over time, to become better and better suited to their environment — sometimes even reaching theoretical limits, as in the aforementioned sensitivity of the eye. Hemoglobin (the key ingredient in red blood cells) is exquisitely adapted to the task of transporting oxygen, tuned by slight variations in different species such that it can load and unload its oxygen cargo in a way optimally suited to the prevailing air pressure — one method for creatures that dwell at sea level, another for a species like the bar-headed goose, an inhabitant of the upper reaches of the Himalayas. From the biochemistry of hemoglobin to the intricate focusing systems of the eye, there are thousands of ways in which biology comes startlingly close to perfection.

But perfection is clearly not always the way; the possibility of imperfection too becomes apparent when we realize that what evolution traverses is not just a mountain, but a mountain *range.* What is omitted from the usual metaphor is the fact that it is perfectly possible for evolution to get *stuck* on a peak that is *short of the highest conceivable summit,* what is known as a "local maximum." As Dawkins and many others have noted, evolution tends to take small steps.* If

---

*Emphasis on "tends to." Strictly speaking, the steps taken by evolution may be of any size, but dramatic mutations rarely survive, whereas small modifications often keep enough core systems in place to have a fighting chance. As a statistical matter, small changes thus appear to have a disproportionately large influence on evolution.

no immediate change leads to an improvement, an organism is likely to stay where it is on the mountain range, even if some distant peak might be better. The kluges I've talked about already — the spine, the inverted retina, and so forth — are examples of just that, of evolution getting stuck on tallish mountains that fall short of the absolute zenith.

In the final analysis, evolution isn't about perfection. It's about what the late Nobel laureate Herb Simon called "satisficing," obtaining an outcome that is good enough. That outcome might be beautiful and elegant, or it might be a kluge. Over time, evolution can lead to both: aspects of biology that are exquisite and aspects of biology that are at best rough-and-ready.

Indeed, sometimes elegance and kluginess coexist, side by side. Highly efficient neurons, for example, are connected to their neighbors by puzzlingly inefficient synaptic gaps, which transform efficient electrical activity into less efficient diffusing chemicals, and these in turn waste heat and lose information. Likewise, the vertebrate eye is, in many respects, tremendously elegant, with its subtle mechanisms for focusing light, adjusting to varied amounts of lighting, and so forth. Though it operates with more sophistication than most digital cameras, it's still hobbled by the backward retina and its attendant blind spot. On the highest peak of evolution, our eyes would work much as they do now, but the retina would face forward (as it does in the octopus), eliminating those blind spots. The human eye is about as good as it could be, given the backward retina, but it could be better — a perfect illustration of how nature occasionally winds up notably short of the highest possible summit.

There are a number of reasons why, at any particular moment, a given creature might have a design that is less than optimal, including random chance (sheer bad luck), rapid environmental change (for example, if there's a major meteor hit, an ice age, or another cataclysmic event, it takes time for evolution to catch up), or the influence that will animate much of this book: history, as encapsulated in our genome. History has a potent — and sometimes detrimental — effect because what can evolve at any given point is heavily constrained by what has evolved before. Just as contemporary political conflicts can in part be traced to the treaties following the world wars, current biology can be traced to the history of earlier creatures. As Darwin put it, all life is the product of "descent with modification"; existing forms are simply altered versions of earlier ones. The human spine, for example, arose not because it was the best possible solution imaginable, but because it was built upon something (the quadruped spine) that already existed.

This gives rise to a notion that I call "evolutionary inertia," borrowing from Newton's law of inertia (an object at rest tends to stay at rest, and an object in motion tends to stay in motion). Evolution tends to work with what is already in place, making modifications rather than starting from scratch.

Evolutionary inertia occurs because new genes must work in concert with old genes and because evolution is driven by the immediate. Gene-bearing creatures either live and reproduce or they don't. Natural selection therefore tends to favor genes that have immediate advantages, discarding other options that might function better in the long term. Thus the process operates a bit like a product manager who needs his product to ship *now*, even if today's cut corners might lead to problems later.

The net result is, as Nobel laureate François Jacob famously put it, that evolution is like a tinkerer "who . . . often without knowing what he is going to produce . . . uses whatever he finds around him, old cardboards, pieces of strings, fragments of wood or metal, to make some kind of workable object . . . [the result is] a patchwork of odd sets pieced together when and where opportunity arose." If necessity is the mother of invention, tinkering is the geeky grandfather of kluge.

In short, evolution often proceeds by piling new systems on top of old ones. The neuroscientist John Allman has captured this idea nicely with an analogy to a power plant he once visited, where at least three layers of technology were in simultaneous use, stacked on top of one another. The recent computer technology operated not directly, but rather by controlling vacuum tubes (perhaps from the 1940s), which in turn controlled still older pneumatic mechanisms that relied on pressurized gases. If the power plant's engineers could afford the luxury of taking the whole system offline, they would no doubt prefer to start over, getting rid of the older systems altogether. But the continuous need for power precludes such an ambitious redesign.

In the same way, living creatures' continuous need to survive and reproduce often precludes evolution from building genuinely optimal systems; evolution can no more take its products offline than the human engineers could, and the consequences are often equally clumsy, with new technologies piled on top of old. The human midbrain, for example, exists literally on top of the ancient hindbrain, and the forebrain is built top of both. The hindbrain, the oldest of the three (dating from at least half a billion years ago), controls respiration, balance, alertness, and other functions that are as critical to a dinosaur as to a human. The midbrain, layered on soon afterward, coordinates visual and auditory reflexes and controls functions such as eye movements. The forebrain, the final division to come online, governs things such as language and decision making, but in ways that often depend on older systems. As any neuroscience textbook will tell you, language relies heavily on Broca's area, a walnut-sized region of the left forebrain, but it too relies on older systems, such as the cerebellum, and ancestral memory systems that are not particularly well suited to the job. Over the course of evolution our brain has become a bit like a palimpsest, an ancient manuscript with layers of text written over it many times, old bits still hiding behind new.

Allman referred to this awkward process, by which new systems are built on top of old ones rather than begun from scratch, as the "progressive overlay of technologies." The end product tends to be a kluge.

Of course, explaining why evolution can produce kluge-like solutions *in general* is not the same thing as showing that the human mind *in particular* is a kluge. But there are two powerful reasons for thinking that it might be: our relatively recent evolution and the nature of our genome.

Consider, first, the short span of human existence and what it might mean. Bacteria have lived on the planet for three billion years, mammals for three hundred million. Humans, in contrast, have been around for, at most, only a few hundred thousand. Language, complex culture, and the capacity for deliberate thought may have emerged only in the past fifty thousand years. By the standards of evolution, that's not a lot of time for debugging, and a long time for the accumulation of prior evolutionary inertia.

Meanwhile, even though your average human makes its living in ways that are pretty different from those of the average monkey, the human genome and primate genomes scarcely differ. Measured nucleotide by nucleotide, the human genome is 98.5 percent identical to that of the chimpanzee. This suggests that the vast majority of our genetic material evolved *in the context of creatures who didn't have language, didn't have culture, and didn't reason deliberately.* This means that the characteristics we hold most dear, the features that most distinctly define us as human beings — language, culture, explicit thought — must have been built on a genetic bedrock *originally adapted for very different purposes.*

Over the course of this book, we'll travel through some of the most important areas of human mental life: memory, belief, choice, language, and pleasure. And in every case, I will show you that kluges abound.

Humans can be brilliant, but they can be stupid too; they can join cults, get addicted to life-ruining drugs, and fall for the claptrap on late-night talk radio. Every one of us is susceptible — not just Joe Sixpack, but doctors, lawyers, and world leaders too, as books like Jerome Groopman's *How Doctors Think* and Barbara Tuchman's *The March of Folly* well attest. Mainstream evolutionary psychology tells us much about how natural selection has led to good solutions, but rather less about why the human mind is so consistently vulnerable to error.

In the pages to come I'll consider why our memory so often fails us, and why we often believe things that aren't true but disbelieve things that are. I'll consider how it is that half of all Americans can believe in ghosts and how almost four million can sincerely believe that they've been abducted by space aliens. I'll look at how we spend (and often waste) our money, why the phenomenon of throwing good money after bad is so widespread, and why we inevitably find meat that is 80 percent lean much more appealing than meat that is 20 percent fat. I'll examine the origins of languages and explain why they are replete with irregularity, inconsistency, and ambiguity — and, for that matter, why a sentence like *People people left left* ties us in knots even though it's only four words long. I'll also look at what makes us happy, and why. It's often been said that pleasure exists to guide the species, but why, for example, do we spend so many hours watching television when it does our genes so little good? And why is mental illness so widespread, affecting, at one time or another, almost half the population? And why on earth *can't* money buy happiness?

Kluge, kluge, kluge. In every case, I'll show that we can best understand our limitations by considering the role of evolutionary inertia in shaping the human mind.

This is not to say that every cognitive quirk is without redeeming value. Optimists often find some solace in even the worst of our mental limitations; if our memory is bad, it is only to protect us from emotional pain; if our language is ambiguous, it is only to enable us to say no without explicitly saying "no."

Well, sort of; there's a difference between being able to exploit ambiguity (say, for purposes of poetry or politeness) and being stuck with it. When our sentences can be misunderstood even when we want them to be clear — or when our memory fails us even when someone's life is at stake (for example, when an eyewitness gives testimony at a criminal trial) — real human cognitive imperfections cry out to be addressed.

I don't mean to chuck the baby along with its bath — or even to suggest that kluges outnumber more beneficial adaptations. The biologist Leslie Orgel once wrote that "Mother Nature is smarter than you are," and most of the time it is. No single individual could ever match what nature has done, and most of nature's designs are sensible, even if they aren't perfect. But it's easy to get carried away with this line of argument. When the philosopher Dan Dennett tells us that "time and again, biologists baffled by some apparently futile or maladroit bit of bad design in nature have eventually come to see that they have underestimated the ingenuity, the sheer brilliance, the depth of insight to be discovered in Mother Nature's creations," he's cheerleading. In an era in which machines can beat humans in intellectual endeavors ranging from chess to statistical analysis, it is possible to contemplate other ways in which physical systems might solve cognitive problems, and nature doesn't always come out on top. Instead of *assuming* that nature is always ingenious, it pays to take each aspect of the mind on its own, to sort the truly sublime from the cases in which nature really could have done better.

Whether kluges outnumber perfections or perfections outnumber kluges, kluges tell us two things that perfections can't. First, they can give special insight into our evolutionary history; when we see perfection, we often can't tell which of many converging factors might have yielded an ideal solution; often it is only by seeing where things went wrong that we can tell how things were built in the first place. Perfection, at least in principle, could be the product of an omniscient, omnipotent designer; imperfections not only challenge that idea but also offer specific forensic clues, a unique opportunity to reconstruct the past and to better understand human nature. As the late Stephen Jay Gould noted, imperfections, "remnants of the past that don't make sense in present terms — the useless, the odd, the peculiar, the incongruous — are the signs of history."

And second, kluges can give us clues into how we can improve ourselves. Whether we are 80 percent perfect or 20 percent perfect (numbers that are really meaningless, since it all depends on how you count), humans do show room for improvement, and kluges can help lead the way. By taking an honest look in the mirror, in recognizing our limitations as well as our strengths, we have a chance to make the most of the noble but imperfect minds we *did* evolve.

2

MEMORY

> Your memory is a monster; you forget — it doesn't. It simply
> files things away. It keeps things for you, or hides things from
> you — and summons them to your recall with a will of its own.
> You think you have a memory; but it has you!
>
> — JOHN IRVING

MEMORY IS, I BELIEVE, the mother of all kluges, the single factor
most responsible for human cognitive idiosyncrasy.

Our memory is both spectacular and a constant source of dis-
appointment: we can recognize photos from our high school year-
books decades later — yet find it impossible to remember what we
had for breakfast yesterday. Our memory is also prone to distortion,
conflation, and simple failure. We can know a word but not be able
to remember it when we need it (think of a word that starts with
a, meaning "a counting machine with beads"),* or we can learn
something valuable (say, how to remove tomato sauce stains) and
promptly forget it. The average high school student spends four years
memorizing dates, names, and places, drill after drill, and yet a significant number of teenagers can't even identify the *century* in which
World War I took place.

I'm one to talk. In my life, I have lost my house keys, my glasses,
my cell phone, and even a passport. I've forgotten where I parked, left
the house without remembering my keys, and on a particularly sad
day, left a leather jacket (containing a second cell phone) on a park
bench. My mother once spent an hour looking for her car in the ga-

---

*The word you're trying to remember is *abacus*.

rage at an unfamiliar airport. A recent *Newsweek* article claims that people typically spend 55 minutes a day "looking for things they know they own but can't find."

Memory can fail people even when their lives are at stake. Skydivers have been known to forget to pull the ripcord to open their parachute (accounting, by one estimate, for approximately 6 percent of skydiving deaths), scuba divers have forgotten to check their oxygen level, and more than a few parents have inadvertently left their babies in locked cars. Pilots have long known that there's only one way to fly: with a checklist, relying on a clipboard to do what human memory can't, which is to keep straight the things that we have do over and over again. (Are the flaps down? Did I check the fuel gauge? Or was that last time?) Without a checklist, it's easy to forget not just the answers but also the questions.

Why, if evolution is usually so good at making things work well, is our memory so hit-or-miss?

The question becomes especially pointed when we compare the fragility of our memory with the robustness of the memory in the average computer. Whereas my Mac can store (and retrieve) my complete address book, the locations of all the countries in Africa, the complete text of every email message I ever sent, and all the photographs I've taken since late 1999 (when I got my first digital camera), not to mention the first 3,000 digits of pi, all in perfect detail, I still struggle with the countries in Africa and can scarcely even remember *whom* I last emailed, let alone exactly what I said. And I never got past the first ten digits of pi (3.1415926535) — even though I was just the sort of nerd who'd try to memorize more.*

Human memory for photographic detail is no better; we can

---

*Which is not to say that no human being could do better. A number of people, far more dedicated to the cause than I ever was, have managed to learn thousands, even tens of thousands, of digits. But it takes *years*. I'd rather go hiking. Still, if you are into that sort of thing, refer to http://www.ludism.org/mentat/PiMemorisation for some basic tips.

recognize the main elements of a photo we've seen before, but studies show that people often don't notice small or even fairly large changes in the background.* And I for one could never *ever* recall the details of a photograph, no matter how long I sat and stared at it beforehand. I can still remember the handful of phone numbers I memorized as a child, when I had loads of free time, but it took me almost a year to learn my wife's cell phone number by heart.

Worse, once we do manage to encode a memory, it's often difficult to revise it. Take, for instance, the trouble I have with the name of my dear colleague Rachel. Five years after she got divorced and reverted to her maiden name (Rachel K.), I still sometimes stumble and refer to her by her former married name (Rachel C.) because the earlier habit is so strong. Whereas computer memory is precise, human memory is in many ways a recalcitrant mess.

Computer memory works well because programmers organize information into what amounts to a giant map: each item is assigned a *specific location,* or "address," in the computer's databanks. With this system, which I will call "postal-code memory," when a computer is prompted to retrieve a particular memory, it simply goes to the relevant address. (A 64-megabyte memory card holds roughly 64 million such addresses, each containing a single "word" made up of a set of eight binary digits.)

Postal-code memory is as powerful as it is simple; used properly, it allows computers to store virtually any information with near-perfect reliability; it also allows a programmer to readily *change* any memory; no referring to Rachel K. as Rachel C. once she's changed her name. It's no exaggeration to say that postal-code memory is a key component of virtually every modern computer.

But not, alas, in humans. Having postal-code memory would

---

*Google for "change blindness" if you've never seen a demonstration; if you haven't seen Derren Brown's "person swap" video on YouTube (www.youtube.com/watch?v= CFaY3YcMg1T), you're missing something special.

have been terrifically useful for us, but evolution never discovered the right part of the mountain range. We humans rarely — if ever — know precisely *where* a piece of information is stored (beyond the extremely vague "somewhere inside the brain"), and our memory evolved according to an entirely different logic.

In lieu of postal-code memory we wound up with what I'll call "contextual memory": we pull things out of our memory by using context, or *clues,* that hints at what we are looking for. It's as if we say to ourselves, every time we need a particular fact, "Um, hello, brain, sorry to bother you, but I need a memory that's about the War of 1812. Got anything for me?" Often, our brain obliges, quickly and accurately yielding precisely the information we want. For instance, if I ask you to name the director who made the movies *E.T.* and *Schindler's List,* you might well come up with the answer within milliseconds — even though you may not have the foggiest idea *where* in your brain that information was stored.* In general, we pull what we need from memory by using various clues, and when things go well, the detail we need just "pops" into our mind. In this respect, accessing a memory is a bit like breathing — most of it comes naturally.

And what comes to mind most naturally often depends on context. We're more likely to remember what we know about gardening when we are in the garden, more likely to remember what we know about cooking when we are in the kitchen. Context, sometimes for better and sometimes for worse, is one of the most powerful cues affecting our memory.

Contextual memory has a *very* long evolutionary history; it's found not just in humans, but also in apes and monkeys, rats and mice, and even spiders and snails. Scientists picked up the first hints of the power of contextual cues almost a hundred years ago, in 1917, when Harvey Carr, a student of the famous behaviorist psychologist John Watson, was running a fairly routine study that involved training rats to run in a maze. Along the way, Carr discovered that the rats

---

*I'm speaking, of course, of Steven Spielberg.

were quite sensitive to factors that had *nothing to do with the maze* it-self. A rat that was trained in a room with electric light, for example, would run the maze better when tested in a room with electric light rather than natural light. The context in which the rat was tested — that is, the environment to which it had grown accustomed — af-fected its memory of how to run in the maze, even though lighting was not germane to the task. It has since become clear that just about every biological creature uses context, relevant or not, as a major guiding force in accessing memory.

Contextual memory may have evolved as a hack, a crude way of compensating for the fact that nature couldn't work out a proper postal-code system for accessing stored information, but there are still some obvious virtues in the system we do have. For one thing, instead of treating all memories equally, as a computer might do, context-dependent memory *prioritizes*, bringing most quickly to mind things that are common, things that we've needed recently, and things that have previously been relevant in situations that are similar to our current circumstances — exactly the sort of informa-tion that we tend to *need* the most. For another thing, context-dependent memories can be searched rapidly in parallel, and as such may represent a good way of compensating for the fact that neurons are millions of times slower than the memory chips used by digital computers. What's more, we (unlike computers) don't have to keep track of the details of our own internal hardware; most of the time, finding what we need in our memory becomes a matter of asking ourself the right question, not identifying a particular set of brain cells.*

Nobody knows for sure how this works, but my best guess is that each of our brain's memories acts *autonomously*, on its own, in re-

---

*The trick — the same goes for using search engines — is to employ as many distinc-tive cues as possible, thereby eliciting fewer and fewer extraneous memories. The more specific your cues are ("where I left my car last night when the street was full" rather than "where I left my car"), the greater the chance that you'll zero in on the fact you need.

sponse to whatever queries it might match, thereby eliminating the need for a central agent to keep a map of memory storage locations. Of course, when you rely on matches rather than specific locations that are known in advance, there's no guarantee that the *right memory* will respond; the fewer the cues you provide, the more "hits" your memory will serve up, and as a consequence the memory that you actually want may get buried among those that you don't want.

Contextual memory has its price, and that price is reliability. Because human memory is so thoroughly driven by cues, rather than location in the brain, we can easily get confused. The reason I can't remember what I had for breakfast yesterday is that yesterday's breakfast is too easily confused with that of the day before, and the day before that. Was it yogurt Tuesday, waffles Wednesday, or the other way around? There are too many Tuesdays, too many Wednesdays, and too many near-identical waffles for a cue-driven system to keep straight. (Ditto for any pilot foolish enough to rely on memory instead of a checklist — one takeoff would blur together with the next. Sooner or later the landing gear would be forgotten.)

Whenever context changes, there's a chance of a problem. I, for example, recently found myself at a party where I was awestruck by the sudden appearance of the luminescent and brilliantly talented actress who played the role of Claire Fisher in the television show *Six Feet Under.* I thought it would be fun to introduce myself. Ordinarily, I probably would have had little trouble remembering her name — I'd seen it in the credits dozens of times, but at that moment I drew a total blank. By the time I got a friend to remind me of her name, the actress was already leaving; I had missed my chance. In hindsight, it's perfectly clear why I couldn't remember her name: the context was all wrong. I was used to seeing the person in question on TV, in character, in a fictional show set in Los Angeles, not in real life, in New York, in the company of the mutual acquaintances who had brought me to the party. In the memory of a human being, context is all, and sometimes, as in this instance, context works against us.

■

Context exerts its powerful effect — sometimes helping us, sometimes not — in part by "priming" the pump of our memory; when I hear the word *doctor,* it becomes easier to recognize the word *nurse.* Had someone said "Lauren" (the first name of the actor in question), I probably could have instantly come up with her last name (Ambrose), but without the right cue, I could only draw a blank.

The thing about context is that it is always with us — even when it's not really relevant to what we are trying to remember. Carr's experiment with rats, for instance, has a parallel with humans in a remarkable experiment with scuba divers. The divers were asked to memorize a list of words while underwater. Like the rats that needed electric light to perform well, the scuba divers were better at remembering the words they studied underwater when they were *tested* underwater (relative to being tested on land) — a fact that strikes this landlubber as truly amazing. Just about every time we remember *anything,* context looms in the background.*

This is not always a good thing. As Merlin Mann of the blog "43 folders" put it, the time when we tend to notice that we need toilet paper tends not to be the moment when we are in a position to *buy* it. Relying on context works fine if the circumstance in which we need some bit of information matches the circumstance in which we first stored it — but it becomes problematic when there is a mismatch between the original circumstance in which we've learned something and the context in which we later need to remember it.

Another consequence of contextual memory is the fact that nearly every bit of information that we hear (or see, touch, taste, or smell), like it or not, triggers some further set of memories — often in ways that float beneath our awareness. The novelist Marcel Proust, who coined the term "involuntary memory," got part of the idea — the

---

*Similarly, if you study while stoned, you might as well take the test while stoned. Or so I have been told.

reminiscences in Proust's famous (and lengthy) novel *Remembrance of Things Past* were all triggered by a single, consciously recognized combination of taste and smell.

But the reality of automatic, unconscious memory exceeds even that which Proust imagined; emotionally significant smells are only the tip of an astonishing iceberg. Take, for example, an ingenious study run by a former colleague of mine, John Bargh, when he was at New York University. His subjects, all undergraduates, were asked to unscramble a series of sentences. Quietly embedded within the scrambled lists were words related to a common theme, such as *old, wise, forgetful,* and *Florida,* designed to elicit the concept of the elderly. The subjects did as they were told, diligently making their way through the task. The real experiment, however, didn't begin until afterward, when Bargh surreptitiously videotaped subjects as they departed after the test, walking to an elevator down the hall. Remarkably, the words people read affected their *walking speed.* The subjects all presumably had places to go and people to see, but those who unscrambled words like *retired* and *Florida* ambled more slowly than those who hadn't.

Another lab studied people as they played a trivia game. Those briefly primed by terms like *professor* or *intelligent* outperformed those prepped with less lofty expressions, such as *soccer hooligans* and *stupid.* All the trash-talking that basketball players do might be more effective than we imagine.

At first, these studies may seem like mere fun and games — stupid pet tricks for humans — but the real-life consequences of priming can be serious. For example, priming can lead minority groups to do worse when cultural stereotypes are made especially salient, and, other things being equal, negative racial stereotypes tend to be primed automatically even in well-intentioned people who report feeling "exactly the same" about whites and blacks. Likewise, priming may reinforce depression, because being in a bad mood primes a person to think about negative things, and this in turn furthers depression. The context-driven nature of memory may also play a role in leading depressed people to seek out depressive activities, such as drinking or listening to songs of lost love, which presumably deepens the gloom as well. So much for intelligent design.

Anchoring our memories in terms of context and cues, rather than specific pre-identified locations, leads to another problem: our memories often blur together. In the first instance, this means that something I learn now can easily interfere with something I knew before: today's strawberry yogurt can obscure yesterday's raspberry. Conversely, something I already know, or once knew, can interfere with something new, as in my trouble with acclimating to Rachel K.'s change in surname.

Ultimately, interference can lead to something even worse: false memories. Some of the first direct scientific evidence to establish the human vulnerability to false memories came from a now classic cognitive-psychology study in which people were asked to memorize a series of random dot patterns like these:

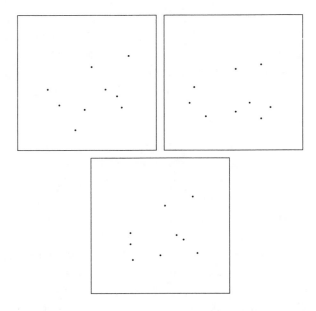

Later, the experimenters showed various dot patterns to the same subjects and asked whether they had seen certain ones before. People were often tricked by this next one, claiming they had seen it when in fact it is a new pattern, a sort of composite of the ones viewed previously.

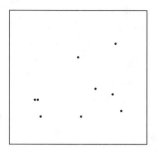

We now know that these sorts of "false alarms" are common. Try, for example, to memorize the following list of words: *bed, rest, awake, tired, dream, wake, snooze, blanket, doze, slumber, snore, nap, peace, yawn, drowsy, nurse, sick, lawyer, medicine, health, hospital, dentist, physician, ill, patient, office, stethoscope, surgeon, clinic, cure.*

If you're like most people, you'll surely remember the *categories* of words I've just asked you to memorize, but you'll probably find yourself fuzzy on the details. Do you recall the word *dream* or *sleep* (or both, or neither?), *snooze* or *tired* (or both, or neither)? How about *doctor* or *dentist*? Experimental data show that most people are easily flummoxed, frequently falling for words they didn't see (such as *doctor*). The same thing appears to happen even with so-called flashbulb memories, which capture events of considerable importance, like 9/11 or the fall of the Berlin Wall. As time passes, it becomes harder and harder to keep particular memories straight, even though we continue to believe, sometimes with great confidence, that they are accurate. Sadly, confidence is no measure of accuracy.

■

For most species, most of the time, remembering gist rather than detail is enough. If you are a beaver, you need to know how to build dams, but you don't need to remember where each individual branch is. For most of evolution, the costs and benefits of context-dependent memory worked out fine: fast for gist, poor for detail; so be it.

If you are human, though, things are often different; societies and circumstances sometimes require of us a precision that wasn't demanded of our ancestors. In the courtroom, for example, it's not enough to know that *some guy* committed a crime; we need to know *which guy* did — which is often more than the average human can remember. Yet, until recently, with the rise of DNA evidence, eyewitness testimony has often been treated as the final arbiter; when an honest-looking witness appears confident, juries usually assume that this person speaks the truth.

Such trust is almost certainly misplaced — not because honest people lie, but because even the most honorable witness is just human — saddled with contextually driven memory. Oodles of evidence for this comes from the lab of the psychologist Elizabeth Loftus. In a typical study, Loftus shows her subjects a film of a car accident and asks them afterward what happened. Distortion and interference rule the day. For example, in one experiment, Loftus showed people slides of a car running a stop sign. Subjects who later heard mention of a yield sign would often *blend* what they saw with what they heard and misremember the car as driving past a yield sign rather than a stop sign.

In another experiment, Loftus asked several different groups of subjects (all of whom had seen a film of another car accident) slightly different questions, such as *How fast were the cars going when they hit each other?* or *How fast were the cars going when they smashed into each other?* All that varied from one version to the next was the final verb (*hit, smashed, contacted,* and so forth). Yet this slight difference in wording was enough to affect people's memory: subjects who heard verbs like *smashed* estimated the crash as occurring at 40.8

miles per hour, a significantly greater speed than that reported by those who heard verbs with milder connotations, like *hit* (34.0) and *contacted* (31.8). The word *smashed* cues different memories than *hit*, subtly influencing people's estimates.

Both studies confirm what most lawyers already know: questions can "lead witnesses." This research also makes clear just how unreliable memory can be. As far as we can tell, this pattern holds just as strongly outside the lab. One recent real-world study, admittedly small, concerned people who had been wrongly imprisoned (and were subsequently cleared on the basis of DNA tests). Over 90 percent of their convictions had hinged on faulty eyewitness testimony.

When we consider the evolutionary origins of memory, we can start to understand this problem. Eyewitness testimony is unreliable because our memories are stored in bits and pieces; without a proper system for locating or keeping them together, context affects how well we retrieve them. Expecting human memory to have the fidelity of a video recorder (as juries often do) is patently unrealistic. Memories related to accidents and crimes are, like all memories, vulnerable to distortion.

A memorable line from George Orwell's novel *1984* states that "Oceania had always been at war with Eurasia" — the irony being, of course, that until recently (in the time frame of the book) Oceania had *not* in fact been at war with Eurasia. ("As Winston well knew, it was only four years since Oceania had been at war with Eastasia and in alliance with Eurasia.") The dictators of *1984* manipulate the masses by revising history. This idea is, of course, essential to the book, but when I read it as a smug teenager, I found the whole thing implausible: wouldn't people *remember* that the battle lines only recently had been redrawn? Who was fooling whom?

Now I realize that Orwell's conceit wasn't so far-fetched. All memories — even those concerning our *own* history — are constantly being revised. Every time we access a memory, it becomes "labile," subject to change, and this seems to be true even for memories that seem especially important and firmly established, such as those of political events or our own experiences.

A good, scientifically well documented illustration of how vulnerable autobiographical memory can be took place in 1992, courtesy of the ever-mercurial Ross Perot, an iconoclastic billionaire from Texas who ran for president as an independent candidate. Perot initially attracted a strong following, but suddenly, under fire, he withdrew from the race. At that point an enterprising psychologist named Linda Levine asked Perot followers how they felt about his withdrawal from the campaign. When Perot subsequently reentered the race, Levine had an unanticipated chance to collect follow-up data. Soon after election day, Levine asked people whom they voted for in the end, and how they felt about Perot earlier in the campaign, at the point when he had dropped out. Levine found that people's memory of *their own feelings* shifted. Those who returned to Perot when Perot reentered the race tended to whitewash their negative memories of his withdrawal, forgetting how betrayed they had felt, while people who moved on from Perot and ultimately voted for another candidate whitewashed their *positive* memories of him, as if they had never intended to vote for him in the first place. Orwell would have been proud.*

Distortion and interference are just the tip of the iceberg. Any number of things would be a whole lot easier if evolution had simply vested us with postal-code memory. Take, for example, the seemingly trivial task of remembering where you last put your house keys. Nine

---

*Several other studies have pointed to the same conclusion — we all tend to be historical revisionists, with a surprisingly dodgy memory of our own prior attitudes. My own personal favorite is an article called "From Chump to Champ" — it describes the one instance in which we all are happy to endure ostensibly negative memories about our own past: when it helps paint us, Rocky-style, as triumphing over adversity.

times out of ten you may get it right, but if you should leave your keys in an atypical spot, all bets are off. An engineer would simply assign a particular memory location (known as a "buffer") to the geographical coordinates of your keys, update the value whenever you moved them, and voilà: you would never need to search the pockets of the pants you wore yesterday or find yourself locked out of your own home.

Alas, precisely because we can't access memories by exact location, we can't straightforwardly update specific memories, and we can't readily "erase" information about where we put our keys in the past. When we place them somewhere other than their usual spot, recency (their most recent location) and frequency (where they're usually placed) come into conflict, and we may well forget where the keys are. The same problem crops up when we try to remember where we last put our car, our wallet, our phone; it's simply part of human life. Lacking proper buffers, our memory banks are a bit like a shoebox full of disorganized photographs: recent photos tend *on average* to be closer to the top, but this is not guaranteed. This shoebox-like system is fine when we want to remember some general concept (say, reliable locations for obtaining food) — in which case, remembering *any* experience, be it from yesterday or a year ago, might do. But it's a lousy system for remembering particular, precise bits of information.

The same sort of conflict between recency and frequency explains the near-universal human experience of leaving work with the intention of buying groceries, only to wind up at home, having completely forgotten to stop at the grocery store. The behavior that is common practice (driving home) trumps the recent goal (our spouse's request that we pick up some milk).

Preventing this sort of cognitive autopilot should have been easy. As any properly trained computer scientist will tell you, driving home and getting groceries are goals, and goals belong on a *stack*. A computer does one thing, then a user presses a key and the first goal (analogous to driving home) is temporarily interrupted by a new goal (getting groceries); the new goal is placed on top of the stack (it becomes top priority), until, when it is completed, it is removed from the stack, returning the old goal to the top. Any number of goals can then be pursued in precisely the right priority sequence. No such luck for us human beings.

Or consider another common quirk of human memory: the fact that our memory for *what* happened is rarely matched by memory for *when* it occurred. Whereas computers and videotapes can pinpoint events to the second (when a particular movie was recorded or particular file was modified), we're often lucky if we can guess the year in which something happened, even if, say, it was in the headlines for months. Most people my age, for example, were inundated a few years ago with a rather sordid story involving two Olympic figure skaters; the ex-husband of one skater hired a goon to whack the other skater on the knee, in order to ruin the latter skater's chance at a medal. It's just the sort of thing the media love, and for nearly six months the story was unavoidable. But if today I asked the average person *when* it happened, I suspect he or she would have difficulty recalling the year, let alone the specific month.*

For something that happened fairly recently, we can get around the problem by using a simple rule of thumb: the more recent the event, the more vivid the memory. But this vividness has its limits: events that have receded more than a couple of months into the past tend to blur together, frequently leaving us chronologically challenged. For example, when regular viewers of the weekly TV news

---

*The principals involved were Tonya Harding, Jeff Gillooly (her ex-husband), and Nancy Kerrigan; Gillooly's hired goon went after Kerrigan's knee on January 6, 1994. Bonus question: When did the Rwandan genocide begin? Answer: April of that same year, three months after the onset of Tonyagate, which was still a big enough story to obscure the news from Rwanda. In the words of the UN commander on the scene, Roméo Dallaire, "During the 100 days of the Rwanda genocide, there was more coverage of Tonya Harding by ABC, CBS, and NBC than of the genocide itself."

program *60 Minutes* were asked to recall when a series of stories aired, viewers could readily distinguish a story presented two months earlier from a story shown only a week before. But stories presented further in the past — say, two years versus four — all faded into an indistinct muddle.

Of course, there is always another workaround. Instead of simply trying to *recall* when something happened, we can try to *infer* this information. By a process known as "reconstruction," we work backward, correlating an event of uncertain date with chronological landmarks that we're sure of. To take another example ripped from the headlines, if I asked you to name the year in which O. J. Simpson was tried for murder, you'd probably have to guesstimate. As vivid as the proceedings were then, they are now (for me, anyway) beginning to get a bit hazy. Unless you are a trivia buff, you probably can't remember exactly when the trial happened. Instead, you might reason that it took place before the Monica Lewinsky scandal but after Bill Clinton took office, or that it occurred before you met your significant other but after you went to college. Reconstruction is, to be sure, better than nothing, but compared to a simple time/date stamp, it's incredibly clumsy.

A kindred problem is reminiscent of the sixth question every reporter must ask. Not *who, what, when, where,* or *why,* but *how,* as in *How do I know it? What are my sources? Where did I see that somewhat frightening article about the Bush administration's desire to invade Iran? Was it in* The New Yorker? *Or the* Economist? *Or was it just some paranoid but entertaining blog?* For obvious reasons, cognitive psychologists call this sort of memory "source memory." And source memory, like our memory for times and dates, is, for want of a proper postal code, often remarkably poor. One psychologist, for example, asked a group of test subjects to read aloud a list of random names (such as Sebastian Weisdorf). Twenty-four hours later he asked them to read a second list of names and to identify which ones belonged to famous people and which didn't. Some were in fact the names of celebrities, and some were made up; the interesting thing is that some were *made-up names drawn from the first list.* If people had good source memory, they would have spotted the ruse. Instead, most subjects knew they had seen a particular name before, but they had no idea where. Recognizing a name like Sebastian Weisdorf but not recalling where they'd seen it, people mistook Weisdorf for the name of a bona fide celebrity whom they just couldn't place. The same thing happens, with bigger stakes, when voters forget whether they heard some political rumor on *Letterman* or read it in the *New York Times.*

The workaround by which we "reconstruct" memory for dates and times is but one example of the many clumsy techniques that humans use to cope with the lack of postal-code memory. If you Google for "memory tricks," you'll find dozens more.

Take for example, the ancient "method of loci." If you have a long list of words to remember, you can associate each one with a specific room in a familiar large building: the first word with the vestibule, the second word with the living room, the third word with the dining room, the fourth with the kitchen, and so forth. This trick, which is used in adapted form by all the world's leading mnemonists, works pretty well, since each room provides a different context for memory retrieval — but it's still little more than a Band-Aid, one more solution we shouldn't need in the first place.

Another classical approach, so prominent in rap music, is to use rhyme and meter as an aid to memorization. Homer had his hexameter, Tom Lehrer had his song "The Elements" ("There's antimony, arsenic, aluminum, selenium, / And hydrogen and oxygen and nitrogen and rhenium . . ."), and the band They Might Be Giants have their cover of "Why Does the Sun Shine? (The Sun Is a Mass of Incandescent Gas)."

Actors often take these mnemonic devices one step further. Not only do they remind themselves of their next lines by using cues of rhythm, syntax, and rhyme; they also focus on their character's motivations and actions, as well as those of other characters. Ideally, this takes place automatically. In the words of the actor Michael Caine, the goal is to get immersed in the story, rather than worry about specific lines. "You must be able to stand there *not* thinking of that line. You take it off the other actor's face." Some performers can do this rather well; others struggle with it (or rely on cue cards). The point is, memorizing lines will never be as easy for us as it would be for a computer. We retrieve memorized information not by reading files from a specific sector of the hard drive but by cobbling together as many clues as possible — and hoping for the best.

Even the oldest standby — simple rehearsal, repeating something over and over — is a bit of clumsiness that shouldn't be necessary. Rote memorization works somewhat well because it exploits the brain's attachment to memories based on frequently occurring events, but here too the solution is hardly elegant. An ideal memory system would capture information in a single exposure, so we wouldn't have to waste time with flash cards or lengthy memorization sessions. (Yes, I've heard the rumors about the existence of photographic memory, but no, I've never seen a well-documented case.)

There's nothing wrong with mnemonics and no end to the possibilities; any cue can help. But when they fail, we can rely on a different sort of solution — arranging our life to accommodate the limits of our memory. I, for example, have learned through long experience that the only way to deal with my congenital absent-mindedness is to develop habits that reduce the demands on my memory. I always put my keys in the same place, position anything I need to bring to work by the front door, and so forth. To a forgetful guy like me, a PalmPilot is a godsend. But the fact that we can patch together solutions doesn't mean that our mental mechanisms are well engineered; it is a symptom of the opposite condition. It is only the clumsiness of human memory that necessitates these tricks in the first place.

Given the liabilities of our contextual memory, it's natural to ask whether its benefits (speed, for example) outweigh the costs. I think not, and not just because the costs are so high, but because it is possible in principle to have the benefits without the costs. The proof is Google (not to mention a dozen other search engines). Search engines start with an underlying substrate of postal-code memory (the well-mapped information they can tap into) and build contextual memory on top. The postal-code foundation guarantees reliability, while the context on top hints at which memories are most likely needed at a given moment. If evolution had *started* with a system of memory organized by location, I bet that's exactly what we'd have, and the advantages would be considerable. But our ancestors never made it to that part of the cognitive mountain; once evolution stumbled upon contextual memory, it never wandered far enough away to find another considerably higher peak. As a result, when we need precise, reliable memories, all we can do is fake it — kluging a poor man's approximation of postal-code memory onto a substrate that doesn't genuinely provide for it.

In the final analysis, we would be nowhere without memory; as Steven Pinker once wrote, "To a very great extent, our memories are ourselves." Yet memory is arguably the mind's original sin. So much is built on it, and yet it is, especially in comparison to computer memory, wildly unreliable.

In no small part this is because we evolved not as computers but as actors, in the original sense of the word: as organisms that act, entities that perceive the world and behave in response to it. And that led to a memory system attuned to speed more than reliability. In many circumstances, especially those requiring snap decisions, recency, frequency, and context are powerful tools for mediating memory. For our ancestors, who lived almost entirely in the here and now (as virtually all nonhuman life forms still do), quick access to contextually relevant memories of recent events or frequently occurring ones helped navigate the challenges of seeking food or avoiding danger. Likewise, for a rat or a monkey, it is often enough to remember related general information. Concerns about misattribution or bias in courtroom testimony simply don't apply.

But today, courts, employers, and many other facets of everyday life make demands that our pre-hominid predecessors rarely faced, requiring us to remember specific details, such as where we *last* put our keys (rather than where we tend, in general, to put them), where we've gotten particular information, and who told us what, and when.

To be sure, there will always be those who see our limits as virtues. The memory expert Henry Roediger, for example, has implied that memory errors are the price we pay in order to make inferences. The Harvard psychologist Dan Schacter, meanwhile, has argued that the fractured nature of memory prepares us for the future: "A memory that works by piecing together bits of the past may be better suited to simulating future events than one that is a store of perfect records." Another common suggestion is that we're better off because we can't remember certain things, as if faulty memory would spare us from pain.

These ideas sound nice on the surface, but I don't see any evidence to support them. The notion that the routine failures of human memory convey some sort of benefit misses an important point: the things that we have trouble remembering *aren't* the things we'd like to forget. It's easy to glibly imagine some kind of optimal state wherein we'd remember only happy thoughts, a bit like Dorothy at the end of *The Wizard of Oz*. But the truth is that we generally can't — contrary to Freud — repress memories that we find painful, and we don't automatically forget them either. What we remember isn't a function of what we *want* to remember, and what we forget isn't a matter of what we *want* to forget; any war veteran or Holocaust survivor could tell you that. What we remember and what we forget are a function of context, frequency, and recency, not a means of attaining inner peace. It's possible to imagine a robot that could automatically expunge all unpleasant memories, but we humans are just not built that way.

Similarly, there is no *logical* relation between having a capacity to make inferences and having a memory that is prone to errors. In principle, it is entirely possible to have both perfect records of past events *and* a capacity to make inferences about the future. That's exactly how computer-based weather-forecasting systems work, for example; they extrapolate the future from a reliable set of data about the past. Degrading the quality of their memory wouldn't improve their predictions, but rather it would undermine them. And there's no evidence that people with an especially distortion-prone memory are happier than the rest of us, no evidence that they make better inferences or have an edge at predicting the future. If anything, the data suggest the opposite, since having an above-average memory is well correlated with general intelligence.

None of which is to say that there aren't compensations. We can, for example, have a great deal of fun with what Freud called "free associations"; it's entertaining to follow the chains of our memories, and we can put that to use in literature and poetry. If connecting trains of thought with chains of ought tickles your fancy, by all means, enjoy! But would we really and truly be better off if our memory was less reliable and more prone to distortion? It's one thing to make lemonade out of lemons, another to proclaim that lemons are what you'd hope for in the first place.

In the final analysis, the fact that our ability to make inferences is built on rapid but unreliable contextual memory isn't some optimal tradeoff. It's just a fact of history: the brain circuits that allow us to make inferences make do with distortion-prone memory because that's all evolution had to work with. To build a truly reliable memory, fit for the requirements of human deliberate reasoning, evolution would have had to start over. And, despite its power and elegance, that's the one thing evolution just can't do.

3

BELIEF

Alice laughed: "There's no use trying," she said; "one can't believe impossible things."

"I daresay you haven't had much practice," said the Queen. "When I was younger, I always did it for half an hour a day. Why, sometimes I've believed as many as six impossible things before breakfast."

— LEWIS CARROLL, *Alice's Adventures in Wonderland*

"YOU HAVE A NEED for other people to like and admire you, and yet you tend to be critical of yourself. While you have some personality weaknesses, you are generally able to compensate for them. You have considerable unused capacity that you have not turned to your advantage. Disciplined and self-controlled on the outside, you tend to be worrisome and insecure on the inside."

Would you believe me if I told you that I wrote that description just for you? It's actually a pastiche of horoscopes, constructed by a psychologist named Bertram Forer. Forer's point was that we have a tendency to read too much into bland generalities, believing that they are (specifically) about us — even when they aren't. Worse, we are even more prone to fall victim to this sort of trap if the bland description includes a few positive traits. Televangelists and late-night infomercials prey upon us in the same way — working hard to sound as if they are speaking to the individual listener rather than a crowd. As a species, we're only too ready to be fooled. This chapter is, in essence, an investigation of why.

The capacity to hold explicit beliefs that we can talk about, evaluate, and reflect upon is, like language, a recently evolved innovation — ubiquitous in humans, rare or perhaps absent in most other species.* And what is recent is rarely fully debugged. Instead of an objective machine for discovering and encoding Truth with a capital *T,* our human capacity for belief is haphazard, scarred by evolution and contaminated by emotions, moods, desires, goals, and simple self-interest — and surprisingly vulnerable to the idiosyncrasies of memory. Moreover, evolution has left us distinctly gullible, which smacks more of evolutionary shortcut than good engineering. All told, though the systems that underlie our capacity for belief are powerful, they are also subject to superstition, manipulation, and fallacy. This is not trivial stuff: beliefs, and the imperfect neural tools we use to evaluate them, can lead to family conflicts, religious disputes, and even war.

In principle, an organism that trafficked in beliefs ought to have a firm grasp on the origins of its beliefs and how strongly the evidence supports them. Does my belief that Colgate is a good brand of toothpaste derive from (1) my analysis of a double-blind test conducted and published by *Consumer Reports,* (2) my enjoyment of Colgate's commercials, or (3) my own comparisons of Colgate against the other "leading brands"? I *should* be able to tell you, but I can't.

---

*Animals often behave as if they too have beliefs, but scientific and philosophical opinion remains divided as to whether they really do. My interest here is the sort of belief that we humans can articulate, such as "On rainy days, it is good to carry an umbrella" or "Haste makes waste." Such nuggets of conventional wisdom aren't necessarily true (if you accept "Absence makes the heart grow fonder," then what about "Out of sight, out of mind"?), but they differ from the more implicit "beliefs" of our sensorimotor system, which we cannot articulate. For example, our sensorimotor system behaves as if it believes that a certain amount of force is sufficient to lift our legs over a curb, but nonphysicists would be hard pressed to say how much force is actually required.) I strongly suspect that many animals have this sort of implicit beliefs, but my working assumption is that beliefs of the kind that we can articulate, judge, and reflect upon are restricted to humans and, at most, a handful of other species.

Because evolution built belief mainly out of off-the-shelf components that evolved for other purposes, we often lose track of where our beliefs come from — if we ever knew — and even worse, we are often completely unaware of how much we are influenced by irrelevant information.

Take, for example, the fact that students rate better-looking professors as teaching better classes. If we have positive feelings toward a given person in one respect, we tend to automatically generalize that positive regard to other traits, an illustration of what is known in psychology as the "halo effect." The opposite applies too: see one negative characteristic, and you expect all of an individual's traits to be negative, a sort of "pitchfork effect." Take, for example, the truly sad study in which people were shown pictures of one of two children, one more attractive, the other less so. The subjects were then told that the child, let's call him Junior, had just thrown a snowball, with a rock inside it, at another child; the test subjects then were asked to interpret the boy's behavior. People who saw the unattractive picture characterized Junior as a thug, perhaps headed to reform school; those shown the more attractive picture delivered judgments that were rather more mild, suggesting, for example, that Junior was merely "having a bad day." Study after study has shown that attractive people get better breaks in job interviews, promotions, admissions interviews, and so on, each one an illustration of how aesthetics creates noise in the channel of belief.

In the same vein, we are more likely to vote for candidates who (physically) "look more competent" than the others. And, as advertisers know all too well, we are more likely to buy a particular brand of beer if we see an attractive person drinking it, more likely to want a pair of sneakers if we see a successful athlete like Michael Jordan wearing them. And though it may be irrational for a bunch of teenagers to buy a particular brand of sneakers so they can "be like Mike," the halo effect, ironically, makes it entirely rational for Nike to spend millions of dollars to secure His Airness's endorsement. And, in a particularly shocking recent study, children of ages three to five gave have higher ratings to foods like carrots, milk, and apple juice if they came in McDonald's packaging. Books and covers, carrots and Styrofoam packaging. We are born to be suckered.

The halo effect (and its devilish opposite) is really just a special case of a more general phenomenon: just about anything that hangs around in our mind, even a stray word or two, can influence how we perceive the world and what we believe. Take, for example, what happens if I ask you to memorize this list of words: *furniture, self-confident, corner, adventuresome, chair, table, independent,* and *television.* (Got that? What follows is more fun if you really do try to memorize the list.)

Now read the following sketch, about a man named Donald:

Donald spent a great amount of his time in search of what he liked to call excitement. He had already climbed Mt. McKinley, shot the Colorado rapids in a kayak, driven in a demolition derby, and piloted a jet-powered boat — without knowing very much about boats. He had risked injury, and even death, a number of times. Now he was in search of new excitement. He was thinking, perhaps, he would do some skydiving or maybe cross the Atlantic in a sailboat.

To test your comprehension, I ask you to sum up Donald in a single word. And the word that pops into your mind is . . . (see the footnote).* Had you memorized a slightly different list, say, *furniture, conceited, corner, reckless, chair, table, aloof, television,* the first word that would have come to mind would likely be different — not *adventuresome,* but *reckless.* Donald may perfectly well be both reckless and adventuresome, but the connotations of each word are very dif-

---

*Due to the effects of memory priming, *adventuresome* is the answer most people give.

ferent — and people tend to pick a characterization that relates to what was already on their mind (in this case, slyly implanted by the memory list). Which is to say that your impression of Donald is swayed by a bit of information (the words in the memory list) that ought to be entirely irrelevant.

Another phenomenon, called the "focusing illusion," shows how easy it is to manipulate people simply by directing their attention to one bit of information or another. In one simple but telling study, college students were asked to answer two questions: "How happy are you with your life in general?" and "How many dates did you have last month?" One group heard the questions in exactly that order, while another heard them in the opposite order, second question first. In the group that heard the question about happiness first, there was almost no correlation between the people's answers; some people who had few dates reported that they were happy, some people with many dates reported that they were sad, and so forth. Flipping the order of the questions, however, put people's focus squarely on romance; suddenly, they could not see their happiness as independent of their love life. People with lots of dates saw themselves as happy, people with few dates viewed themselves as sad. Period. People's judgments in the dates-first condition (but not in the happiness-first condition) were strongly correlated with the number of dates they'd had. This may not surprise you, but it ought to, because it highlights just how malleable our beliefs really are. Even our own internal sense of self can be influenced by what we happen to focus on at a given moment.

The bottom line is that every belief passes through the unpredictable filter of contextual memory. Either we directly recall a belief that we formed earlier, or we calculate what we believe based on whatever memories we happen to bring to mind.

Yet few people realize the extent to which beliefs can be contaminated by vagaries of memory. Take the students who heard the dating question first. They presumably *thought* that they were answering the happiness question as objectively as they could; only an exceptionally self-aware undergraduate would realize that the answer to the second question might be biased by the answer to the first. Which is precisely what makes mental contamination so insidious. Our subjective impression that we are being objective rarely matches the objective reality: no matter how hard we try to be objective, human beliefs, because they are mediated by memory, are inevitably swayed by minutiae that we are only dimly aware of.

From an engineering standpoint, humans would presumably be far better off if evolution had supplemented our contextually driven memory with a way of *systematically* searching our inventory of memories. Just as a pollster's data are most accurate if taken from a representative cross section of a population, a human's beliefs would be soundest if they were based on a balanced set of evidence. But alas, evolution never discovered the statistician's notion of an unbiased sample.

Instead, we routinely take whatever memories are most recent or most easily remembered to be much more important than any other data. Consider, for example, an experience I had recently, driving across country and wondering at what time I'd arrive at the next motel. When traffic was moving well, I'd think to myself, "Wow, I'm driving at 80 miles per hour on the interstate; I'll be there in an hour." When traffic slowed due to construction, I'd say, "Oh no, it'll take me two hours." What I was almost comically unable to do was to take an average across two data points at the same time, and say, "Sometimes the traffic moves well, sometimes it moves poorly. I anticipate a mixture of good and bad, so I bet it will take an hour and a half."

Some of the world's most mundane but common interpersonal friction flows directly from the same failure to reflect on how well our samples represent reality. When we squabble with our spouse or our roommate about whose turn it is to wash the dishes, we probably (without realizing it) are better able to remember the previous times when we, ourself, took care of them (as compared to the times when our roommate or spouse did); after all, our memory is organized to focus primarily on our own experience. And we rarely compensate for that imbalance — so we come to believe we've done more work overall and perhaps end up in a self-righteous huff. Studies show that in virtually any collaborative enterprise, from taking care of a household to writing academic papers with colleagues, the sum of each individual's perceived contribution exceeds the total amount of work done. We cannot remember what other people did as well as we recall what we did ourselves — which leaves everybody (even shirkers!) feeling that others have taken advantage of them. Realizing the limits of our own data sampling might make us all a lot more generous.

Mental contamination is so potent that even entirely irrelevant information can lead us by the nose. In one pioneering experiment, the psychologists Amos Tversky and Daniel Kahneman spun a wheel of fortune, marked with the numbers 1–100, and then asked their subjects a question that had nothing to do with the outcome of spinning the wheel: what percentage of African countries are in the United Nations? Most participants didn't know for sure, so they had to estimate — fair enough. But their estimates were considerably affected by *the number on the wheel*. When the wheel registered 10, a typical response to the UN question was 25 percent, whereas when the wheel came up at 65, a typical response was 45 percent.*

This phenomenon, which has come to be known as "anchoring and adjustment," occurs again and again. Try this one: Add 400 to the last three digits of your cell phone number. When you're done, answer the following question: in what year did Attila the Hun's rampage through Europe finally come to an end? The average guess of

---

*Nobody's ever been able to tell me whether the original question was meant to ask how many of the countries in Africa were in the UN, or how many of the countries in the UN were in Africa. But in a way, it doesn't matter: anchoring is strong enough to apply even when we don't know precisely what the question is.

people whose phone number, plus 400, yields a sum less than 600 was A.D. 629, whereas the average guess of people whose phone number digits plus 400 came in between 1,200 and 1,399 was A.D. 979, 350 years later.*

What's going on here? Why should a phone number or a spin on a wheel of fortune influence a belief about history or the composition of the UN? During the process of anchoring and adjustment, people begin at some arbitrary starting point and keep moving until they find an answer they like. If the number 10 pops up on the wheel, people start by asking themselves, perhaps unconsciously, "Is 10 a plausible answer to the UN question?" If not, they work their way up until they find a value (say, 25) that seems plausible. If 65 comes up, they may head in the opposite direction: "Is 65 a plausible answer? How about 55?" The trouble is, anchoring at a single arbitrarily chosen point can steer us toward answers that are just barely plausible: starting low leads people to the *lowest* plausible answer, but starting high leads them to the *highest* plausible answer. Neither strategy directs people to what might be the most sensible response — one *in the middle of the range of plausible answers*. If you think that the correct answer is somewhere between 25 and 45, why say 25 or 45? You're probably better off guessing 35, but the psychology of anchoring means that people rarely do.†

Anchoring has gotten a considerable amount of attention in psychological literature, but it's by no means the only illustration of how beliefs and judgments can be contaminated by peripheral or even irrelevant information. To take another example, people who are asked to hold a pen between their teeth gently, without letting it

---

*When did Attila actually get routed? A.D. 451.

†If you're aware of the process of anchoring and adjustment, you can see that why it is that during a financial negotiation it's generally better to *make* the opening bid than to respond to it. This phenomenon also explains why, as one recent study showed, supermarkets can sell more cans of soup with signs that say LIMIT 12 PER CUSTOMER rather than LIMIT 4 PER CUSTOMER.

touch their lips, rate cartoons as more enjoyable than do people who hold a pen with pursed lips. Why should that be? You can get a hint if you try following these instructions while looking in a mirror: Hold a pen between your teeth "gently, without letting it touch the lips." Now *look* at the shape of your lips. You'll see that the corners are up-turned, in the position of a smile. And thus, through the force of context-dependent memory, upturned lips tend to automatically lead to happy thoughts.

A similar line of experiments asked people to use their non-dominant hand (the left, for right-handed people) to write down names of celebrities as fast as they could while classifying them into categories (*like, don't like, neutral*). They had to do this while either (1) pressing their dominant hand, palm down, against the top of a ta-ble or (2) pushing their dominant hand, palm upward, against the bottom of a table. Palms-up people listed more positive than negative names, while palms-down people produced more negative names than positive. Why? Palms-up people were positioned in a positive "approach" posture while palms-down people were positioned in an "avoid" posture. The data show that such subtle differences routinely affect our memories and, ultimately, our beliefs.

Another source of contamination is a kind of mental shortcut, the human tendency to believe that what is familiar is good. Take, for ex-ample, an odd phenomenon known as the "mere familiarity" effect: if you ask people to rate things like the characters in Chinese writing, they tend to prefer those that they have seen before to those they haven't. Another study, replicated in at least 12 different languages, showed that people have a surprising attachment to the *letters* found in their own names, preferring words that contain those letters to words that don't. One colleague of mine has even suggested, some-what scandalously, that people may love famous paintings as much for their familiarity as for their beauty.

From the perspective of our ancestors, a bias in favor of the familiar may well have made sense; what great-great-great-grandma knew and didn't kill her was probably a safer bet than what she didn't know — which might do her in. Preference for the familiar may well have been adaptive in our ancestors, selected for in the usual ways: creatures with a taste for the well known may have had more offspring than creatures with too extreme a predilection for novelty. Likewise, our desire for comfort foods, presumably those most familiar to us, seems to increase in times of stress; again, it's easy to imagine an adaptive explanation.

In the domain of aesthetics, there's no real downside to preferring what I'm already used to — it doesn't really matter whether I like this Chinese character better than that one. Likewise, if my love of 1970s disco stems from mere familiarity rather than the exquisite musicianship of Donna Summer, so be it.

But our attachment to the familiar can be problematic too, especially when we don't recognize the extent to which it influences our putatively rational decision making. In fact, the repercussions can take on global significance. For example, people tend to prefer social policies that are already in place to those that are not, even if no well-founded data prove that the current policies are working. Rather than analyze the costs and benefits, people often use this simple heuristic: "If it's in place, it must be working."

One recent study suggested that people will do this even when they have no idea what policies are in place. A team of Israeli researchers decided to take advantage of the many policies and local ordinances that most people know little about. So little, in fact, that the experimenters could easily get the subjects to believe whatever they suggested; the researchers then tested how attached people had become to whatever "truth" they had been led to believe in. For example, subjects were asked to evaluate policies such as the feeding of alley cats — should it be okay, or should it be illegal? The experimenter told half the subjects that alley-cat feeding was currently legal and the other half that it wasn't, and then asked people whether the policy should be changed. Most people favored whatever the current policy was and tended to generate more reasons to favor it over the competing policy. The researchers found similar results with made-up rules about arts-and-crafts instruction. (Should students have five hours of instruction or seven? The current policy is X.) The same sort of love-the-familiar reasoning applies, of course, in the real world, where the stakes are higher, which explains why incumbents are almost always at an advantage in an election. Even recently deceased incumbents have been known to beat their still-living opponents.*

The more we are threatened, the more we tend to cling to the familiar. Just think of the tendency to reach for comfort food. Other things being equal, people under threat tend to become more attached than usual to their own groups, causes, and values. Laboratory studies, for example, have shown that if you make people contemplate their own death ("Jot down, as specifically as you can, what you think will happen to *you* as you physically die . . ."), they tend to be nicer than normal to members of their own religious and ethnic groups, but more negative toward outsiders. Fears of death also tend to polarize people's political and religious beliefs: patriotic Americans who are made aware of their own mortality are more appalled (than patriots in a control group) by the idea of using the American flag as a sieve; devout Christians who are asked to reflect upon their own death are less tolerant of someone using a crucifix as a substitute hammer. (Charities, take note: we also open up our wallets more when we've just thought about death.) Another study has shown that all people tend to become more negative toward minority groups in times of crisis; oddly enough, this holds true not just for members of the majority but even for members of the minority groups themselves.

People may even come to love, or at least accept, systems of gov-

---

*In March 2006, in Sierra Vista, Arizona, Bob Kasun, dead for nine days, won by a margin of nearly three to one.

ernment that profoundly threaten their self-interest. As the psychologist John Jost has noted, "Many people who lived under feudalism, the Crusades, slavery, communism, apartheid, and the Taliban believed that their systems were imperfect but morally defensible and [even sometimes] better than the alternatives they could envision." In short, mental contamination can be very serious business.

Each of these examples of mental contamination — the focusing illusion, the halo effect, anchoring and adjustment, and the familiarity effect — underscores an important distinction that will recur throughout this book: as a rough guide, our thinking can be divided into two streams, one that is fast, automatic, and largely unconscious, and another that is slow, deliberate, and judicious.

The former stream, which I will refer to as the ancestral system, or the reflexive system, seems to do its thing rapidly and automatically, with or without our conscious awareness. The latter stream I will call the deliberative system, because that's what it does: it deliberates, it considers, it chews over the facts — and tries (sometimes successfully, sometimes not) to reason with them.

The reflexive system is clearly older, found in some form in virtually every multicellular organism. It underlies many of our everyday actions, such as the automatic adjustment of our gait as we walk up and down an uneven surface, or our rapid recognition of an old friend. The deliberative system, which consciously considers the logic of our goals and choices, is a lot newer, found in only a handful of species, perhaps only humans.

As best we can tell, the two systems rely on fairly different neural substrates. Some of the reflexive system depends on evolutionarily old brain systems like the cerebellum and basal ganglia (implicated in motor control) and the amygdala (implicated in emotion). The deliberative system, meanwhile, seems to be based primarily in the forebrain, in the prefrontal cortex, which is present — but vastly smaller — in other mammals.

I describe the latter system as "deliberative" rather than, say, rational because there is no guarantee that the deliberative system will deliberate in genuinely rational ways. Although this system can, in principle, be quite clever, it often settles for reasoning that is less than ideal. In this respect, one might think the deliberative system as a bit like the Supreme Court: its decisions may not always seem sensible, but there's always at least an intention to be judicious.

Conversely, the reflexive system shouldn't be presumed irrational; it is certainly more shortsighted than the deliberative system, but it likely wouldn't exist at all if it were completely irrational. Most of the time, it does what it does well, even if (by definition) its decisions are not the product of careful thought. Similarly, although it might seem tempting, I would also caution against equating the reflexive system with emotions. Although many (such as fear) are arguably reflexive, emotions like schadenfreude — the delight one can take in a rival's pain — are not. Moreover, a great deal of the reflexive system has little if anything to do with emotion; when we instinctively grab a railing as we stumble on a staircase, our reflexive system is clearly what kicks in to save us — but it may do so entirely without emotion. The reflexive system (really, perhaps a set of systems) is about making snap judgments based on experience, emotional or otherwise, rather than feelings per se.

Even though the deliberative system is more sophisticated, the latest in evolutionary technology, it has never really gained proper control because it bases its decisions on almost invariably second-hand information, courtesy of the less-than-objective ancestral system. We can reason as carefully as we like, but, as they say in computer science jargon, "garbage in, garbage out." There's no guarantee that the ancestral system will pass along a balanced set of data. Worse, when we are stressed, tired, or distracted, our deliberative system tends to be the first thing to go, leaving us at the mercy of our lower-tech reflexive system — just when we might need our deliberative system the most.

The unconscious influence of our ancestral system is so strong that when our conscious mind tries to get control of the situation, the effort sometimes backfires. For example, in one study, people were put under time pressure and asked to make rapid judgments. Those who were told to (deliberately) suppress sexist thoughts (themselves presumably the product of the ancestral reflexive system) actually became *more likely* than control subjects to have sexist thoughts. Even more pernicious is the fact that as evolution layered reason on top of contextually driven memory, it left us with the *illusion* of objectivity. Evolution gave us the tools to deliberate and reason, but it didn't give us any guarantee that we'd be able to use them without interference. We feel as if our beliefs are based on cold, hard facts, but often they are shaped by our ancestral system in subtle ways that we are not even aware of.

No matter what we humans think about, we tend to pay more attention to stuff that fits in with our beliefs than stuff that might challenge them. Psychologists call this "confirmation bias." When we have embraced a theory, large or small, we tend to be better at noticing evidence that supports it than evidence that might run counter to it.

Consider the quasi-astrological description that opened this chapter. A person who wants to believe in astrology might notice the parts that seem true ("you have a need for other people to like and admire you") and ignore the parts that aren't (maybe from the outside you don't really look so disciplined after all). A person who wishes to believe in horoscopes may notice the one time that their reading seems dead-on and ignore (or rationalize) the thousands of times when their horoscopes are worded so ambiguously that they could mean anything. That's confirmation bias.

Take, for example, an early experiment conducted by the British psychologist Peter Wason. Wason presented his subjects with a triplet of three distinct numbers (for example, 2-4-6) and asked them to guess what rule might have generated their arrangement. Subjects were then asked to create new sequences and received feedback as to whether their new sequences conformed to the rule. A typical subject might guess "4-6-8," be told yes, and proceed to try "8-10-12" and again be told yes; the subject might then conclude that the rule was something like "sequences of three even numbers with two added each time." What most people failed to do, however, was consider potentially *disconfirming* evidence. For example, was 1-3-5 or 1-3-4 a valid sequence? Few subjects bothered to ask; as a consequence, hardly anybody guessed that the actual rule was simply "any sequence of three ascending numbers." Put more generally, people all too often look for cases that confirm their theories rather than consider whether some alternative principle might work better.

In another, later study, less benign, two different groups of people saw a videotape of a child taking an academic test. One group of viewers was led to believe that the child came from a socioeconomically privileged background, the other to believe that the child came from a socioeconomically impoverished background. Those who thought the child was wealthier reported that the child was doing well and performing above grade level; the other group guessed that the child was performing below grade level.

Confirmation bias might be an inevitable consequence of contextually driven memory. Because we retrieve memory not by systematically searching for all relevant data (as computers do) but by finding things that *match*, we can't help but be better at noticing things that confirm the notions we begin with. When you think about the O. J. Simpson murder trial, if you were predisposed to think he was guilty, you're likely to find it easier to remember evidence that pointed toward his guilt (his motive, the DNA evidence, the lack of other plausible suspects) rather than evidence that cast doubt on it (the shoddy police work and that infamous glove that didn't fit).

To consider something *well*, of course, is to evaluate both sides of an argument, but unless we go the extra mile of deliberately forcing ourselves to consider alternatives — not something that comes naturally — we are more prone to recall evidence consistent with an accepted proposition than evidence inconsistent with it. And since we most clearly remember information that seems consistent with our beliefs, it becomes very hard to let those beliefs go, even when they are erroneous.

The same, of course, goes for scientists. The aim of science is to take a balanced approach to evidence, but *scientists* are human beings, and human beings can't help but notice evidence that confirms their own theories. Read any science texts from the past and you will stumble on not only geniuses, but also people who in hindsight seem like crackpots — flat-earthers, alchemists, and so forth. History is not kind to scientists who believed in such fictions, but a realist might recognize that in a species so dependent on memory driven by context, such slip-ups are always a risk.

In 1913 Eleanor Porter wrote one of the more influential children's novels of the twentieth century, *Pollyanna,* a story of a girl who looked on the bright side of every situation. Over time, the name Pollyanna has become a commonly used term with two different connotations. It's used in a positive way to describe eternal optimists and in a negative way to describe people whose optimism exceeds the rational bounds of reality. Pollyanna may have been a fictional character, but there's a little bit of her in all of us, a tendency to perceive the world in positive ways that may or may not match reality. Generals and presidents fight on in wars that can't be won, and scientists retain beliefs in pet theories long after the weight of evidence is stacked against them.

Consider the following study, conducted by the late Ziva Kunda. A group of subjects comes into the lab. They are told they'll be playing a trivia game; before they play, they get to watch someone else, who, they are told, will play either on their team (half the subjects hear this) or on the opposite team (that's what the other half are told). Unbeknownst to the subjects, the game is rigged; the person they're watching proceeds to play a perfect game, getting every question right. The researchers want to know whether each subject is impressed by this. The result is straight out of *Pollyanna:* people who expect to play *with* the perfect-game-playing confederate are impressed; the guy must be great, they think. People who expect to play a*gainst* the confederate are dismissive; they attribute his good performance to luck rather than skill. Same data, different interpretation: both groups of subjects observe someone play a perfect game, but what they make of that observation depends on the role they expect the observed man to play in their own life.

In a similar study, a bunch of college students viewed videos of three people having a conversation; they were asked to judge how likable each of the three was. The subjects were also told (prior to watching the video) that they would be going out on a date with one of those three people (selected at random for each subject). Inevitably, subjects tended to give their highest rating to the person they were told they would be dating — another illustration of how easily our beliefs (in this case, about someone's likability) can be contaminated by what we *wish* to believe. In the words of a musical I loved as a child, Harry Nilsson's *The Point!,* "You see what you want to see, and you hear want you want to hear. Dig?"

Our tendency to accept what we wish to believe (what we are motivated to believe) with much less scrutiny than what we don't want to believe is a bias known as "motivated reasoning," a kind of flip side to confirmation bias. Whereas confirmation bias is an automatic tendency to notice data that fit with our beliefs, motivated reasoning is the complementary tendency to scrutinize ideas more carefully if we don't like them than if we do. Take, for example, a study in which Kunda asked subjects, half men, half women, to read an article claiming that caffeine was risky for women. In line with the notion that our beliefs — and reasoning — are contaminated by motivation, women who were heavy caffeine drinkers were more likely to doubt the conclusion than were women who were light caffeine drinkers; meanwhile, men, who thought they had nothing at stake, exhibited no such effect.

The same thing happens all the time in the real world. Indeed, one of the first scientific illustrations of motivated reasoning was not a laboratory experiment but a clever bit of real-world fieldwork conducted in 1964, just after the publication of the first Surgeon General's report on smoking and lung cancer. The Surgeon General's conclusion — that smoking appears to cause lung cancer — would hardly seem like news today, but at the time it was a huge deal, covered widely by the media. Two enterprising scientists went out and interviewed people, asking them to evaluate the Surgeon General's conclusion. Sure enough, smokers were less persuaded by the report than were nonsmokers, who pretty much accepted what the Surgeon General had to say. Smokers, meanwhile, came up with all kinds of dubious counterarguments: "many smokers live a long time" (which ignored the statistical evidence that was presented), "lots of things are hazardous" (a red herring), "smoking is better than excessive eating or drinking" (again irrelevant), or "smoking is better than being a nervous wreck" (an assertion that was typically not supported by any evidence).

The reality is that we are just not born to reason in balanced ways; even sophisticated undergraduates at elite universities tend to fall prey to this weakness. One famous study, for example, asked students at Stanford University to evaluate a set of studies on the effectiveness of capital punishment. Some of the students had prior beliefs in favor of capital punishment, some against. Students readily found holes in studies that challenged what they believed but often missed equally serious problems with studies that led to conclusions that they were predisposed to agree with.

Put the contamination of belief, confirmation bias, and motivated reasoning together, and you wind up with a species prepared to believe, well, just about anything. Historically, our species has believed in a flat earth (despite evidence to the contrary), ghosts, witches, astrology, animal spirits, and the benefits of self-flagellation and blood-letting. Most of those particular beliefs are, mercifully, gone today, but some people still pay hard-earned money for psychic readings and séances, and even I sometimes hesitate before walking under a ladder. Or, to take a political example, some 18 months after the 2003 invasion of Iraq, 58 percent of people who voted for George W. Bush still believed there were weapons of mass destruction in Iraq, despite the evidence to the contrary.

And then there is President George W. Bush himself, who reportedly believes that he has a personal and direct line of communication with an omniscient being. Which, as far as his getting elected was concerned, was a good thing; according to a February 2007 Pew Research Center survey, 63 percent of Americans would be reluctant to vote for anyone who doesn't believe in God.

To critics like Sam Harris (author of the book *The End of Faith*), that sort of thing seems downright absurd:

> To see how much our culture currently partakes of . . . irratio-nality . . . just substitute the names of your favorite Olympian for "God" wherever this word appears in public discourse. Imagine President Bush addressing the National Prayer Breakfast in these terms: "Behind all of life and all history there is a dedica-tion and a purpose, set by the hand of a just and faithful Zeus." Imagine his speech to Congress (September 20, 2001) containing the sentence "Freedom and fear, justice and cruelty have always been at war and we know that Apollo is not neutral between them."

Religion in particular enjoys the sway that it does in part be-cause people *want* it to be true; among other things, religion gives people a sense that the world is just and that hard work will be re-warded. Such faith provides a sense of purpose and belonging, in both the personal and the cosmic realms; there can be no doubt that the desire to believe contributes to the capacity to do so. But none of that explains how people manage to cling to religious beliefs despite the manifest lack of direct *evidence*.* For that we must turn to the fact that evolution left us with the capacity to fool ourselves into believing what we want to believe. (If we pray and something good happens, we notice it; if nothing happens, we fail to notice the non-coincidence.) Without motivated reasoning and confirmation bias, the world might be a very different place.

As one can see in the study of cigarette smokers, biased reasoning has at least one benefit. It can help protect our self-esteem. (Of course it's not just smokers; I've seen scientists do much the same thing, nit-picking desperately at studies that challenge beliefs to which they're attached.)

The trouble, of course, is that self-deception often costs us down the road. When we fool ourselves with motivated reasoning, we may hold on to beliefs that are misguided or even delusional. They can cause social friction (when we abruptly dismiss the views of others), they can lead to self-destruction (when smokers dismiss the risks of their habit), and they can lead to scientific blunders (when scientists refuse to recognize data challenging their theories).

---

*Some (thankfully not all) of those who believe in creationism rather than evolution appear eager to take just about *any* evidence as further confirmation of their views. One religious news site, for example, took the recent discovery that human DNA is more variable than once thought to "debunk evolution." The argument ran thusly (and I quote verbatim):

> Given that humans are ten times [more] different than one another [than expected], it would seem that a four percentage point difference between the chimpanzee and the human genome could mean hundreds of times differences between each individual human and each individual chimpanzee. And this difference would demolish any reasonable defense of evolution . . . the more scientists find, the more the Bible is proven.

If there had been *less* genetic variation, the argument might well have run, "We are all made in God's image; therefore it's no surprise that our DNA is all the same," but in the quoted passage there seems to be no discernible logic relating the premise (that there is an unexpectedly large amount of genetic variation) to its rather antiscientific conclusion.

When people in power indulge in motivated reasoning, dismissing important signs of their own error, the results can be catastrophic. Such was probably the case, for example, in one of the great blunders in modern military history, in the spring of 1944, when Hitler, on the advice of his leading field marshal, Gerd von Rundstedt, chose to protect Calais rather than Normandy, despite the prescient lobbying of a lesser-ranked general, Erwin Rommel. Von Rundstedt's bad advice, born of undue attachment to his own plans, cost Hitler France, and possibly the entire Western Front.*

Why does motivated reasoning exist in the first place? Here, the problem is not one of evolutionary inertia but a simple lack of foresight. While evolution gave us the gift of deliberate reasoning, it lacked the vision to make sure we used it wisely: nothing forces us to be evenhanded because there was no one there to foresee the dangers inherent in pairing powerful tools of reasoning with the risky temptations of self-deception. In consequence, by leaving it up to our conscious self to decide how *much* to use our mechanism of deliberate reasoning, evolution freed us — for better or for worse — to be as biased as we want to be.

Even when we have little at stake, what we already know — or think we know — often further contaminates our capacity to reason and form new beliefs. Take, for example, the classic form of logic known as the syllogism: a formal deductive argument consisting of major premise, minor premise, and conclusion — as stylized as a sonnet:

> All men are mortal.
> *Socrates was a man.*
> Therefore, Socrates was mortal.

---

*Why didn't von Rundstedt listen? He was too attached to his own strategy, an elaborate but ultimately pointless plan for defending Calais. Hitler, for his part, trusted von Rundstedt so much that he spent the morning of D-Day asleep, evidently untroubled by Rommel's fear that Normandy might be invaded.

Nobody has trouble with this form of logic; we understand the abstract form and realize that it generalizes freely:

All glorks are frum.
Skeezer is a glork.
Therefore, Skeezer is frum.

Presto — a new way for forming beliefs: take what you know (the minor and major premises), insert them into the inferential schema (all X's are Y, Q is an X, therefore Q is a Y), and deduce new beliefs. The beauty of the scheme is the way in which true premises are guaranteed, by the rules of logic, to lead to true conclusions.

The good news is that humans can do this sort of thing at all; the bad news is that, without a lot of training, we don't do it particularly well. If the capacity to reason logically is the product of natural selection, it is also a very recent adaptation with some serious bugs yet to be worked out.

Consider, for example, this syllogism, which has a slight but important difference from the previous one:

All living things need water.
Roses need water.
Therefore, roses are living things.

Is this a valid argument? Focus on the logic, not the *conclusion* per se; we already know that roses are living things. The question is whether the logic is sound, whether the conclusion follows the premises like the night follows the day. Most people think the argument is solid. But look carefully: the statement that all living things need water doesn't preclude the possibility that some *non*living things might need water too. My car's battery, for instance.

The poor logic of the argument becomes clearer if I simply change the words in question:

Premise 1: All insects need oxygen.
Premise 2: Mice need oxygen.
Conclusion: Therefore, mice are insects.

A creature truly noble in reason ought to see, instantaneously, that the rose and mouse arguments follow exactly the same formal structure (all X's need Y, Z's need Y, therefore Z's are X's) and ought to instantly reject all such reasoning as fallacious. But most of us need to see the two syllogisms side by side in order to get it. All too often we suspend a careful analysis of what is *logical* in favor of prior beliefs.

What's going on here? In a system that was superlatively well engineered, belief and the process of drawing inferences (which soon become new beliefs) would be separate, with an iron wall between them; we would be able to distinguish what we had direct evidence for from what we had merely inferred. Instead, in the development of the human mind, evolution took a different path. Long before human beings began to engage in completely explicit, formal forms of logic (like syllogisms), creatures from fish to giraffes were probably making informal inferences, automatically, without a great deal of reflection; if apples are good to eat, pears probably are too. A monkey or a gorilla might make that inference without ever realizing that there *is* such a thing as an inference. Perhaps one reason people are so apt to confuse what they know with what they have merely inferred is that for our ancestors, the two were scarcely different, with much of inference arising automatically as part of belief, rather than via some separate, reflective system.

The capacity to codify the laws of logic — to recognize that *if P, then Q; P; therefore Q* is valid whereas *if P, then Q; Q; therefore P* is not — presumably evolved only recently, perhaps sometime after the arrival of *Homo sapiens*. And by that time, belief and inference were already too richly intertwined to allow the two to ever be fully separate in everyday reasoning. The result is very much a kluge: a perfectly sound system of deliberate reasoning, all too often pointlessly clouded by prejudice and prior belief.

Studies of the brain bear this out: people evaluate syllogisms using two different neural circuits, one more closely associated with logic and spatial reasoning (bilateral parietal), the other more closely associated with prior belief (frontal-temporal). The former (logical and spatial) is effortful, the latter invoked automatically; getting the logic right is difficult.

In fact, truly explicit reasoning via logic probably isn't something that *evolved*, per se, at all. When humans do manage to be rational, in a formal logical sense, it's not because we are built that way, but because we are clever enough to learn the rules of logic (and to recognize their validity, once explained). While all normal human beings acquire language, the ability to use formal logic to acquire and reason about beliefs may be more of a cultural product than an evolutionary one, something made possible by evolution but not guaranteed by it. Formal reason seems to be present, if at all, primarily in literate cultures but difficult to discern in preliterate ones. The Russian psychologist Alexander Luria, for example, went to the mountains of central Asia in the late 1930s and asked the indigenous people to consider the logic of syllogisms like this one: "In a certain town in Siberia all bears are white. Your neighbor went to that town and he saw a bear. What color was that bear?" His respondents just didn't get it; a typical response would be, in essence, "How should I know? Why doesn't the professor go ask the neighbor himself?" Further studies later in the twentieth century essentially confirmed this pattern; people in nonliterate societies generally respond to queries about syllogisms by relying on the facts that they already know, apparently blind to the abstract logical relations that experimenters are inquiring about. This does not mean that people from those societies cannot learn formal logic — in general, at least the children can — but it does show that acquiring an abstract logic is not a natural, automatic phenomenon in the way that acquiring language is. This in turn suggests that formal tools for reasoning about belief are at least as much learned as they are evolved, not (as assumed by proponents of the idea that humanity is innately rational) standard equipment.

Once we decide something is true (for whatever reason), we often make up new reasons for believing it. Consider, for example, a study that I ran some years ago. Half my subjects read a report of a study that showed that good firefighting was correlated with high scores on a measure of risk-taking ability; the other half of the subjects read the opposite: they were told of a study that showed that good firefighting was *negatively* correlated with risk-taking ability, that is, that risk takers made poor firefighters. Each group was then further subdivided. Some people were asked to reflect on what they read, writing down reasons for why the study they read about might have gotten the results it did; others were simply kept busy with a series of difficult geometrical puzzles like those found on an IQ test.

Then, as social psychologists so often do, I pulled the rug out from under my subjects: "Headline, this news just in — the study you read about in the first part of the experiment was a fraud. The scientists who allegedly studied firefighting actually made their data up! What I'd like to know is what you really think — is firefighting really correlated with risk taking?"

Even after I told people that the original study was complete rubbish, people in the subgroups who got a chance to reflect (and create their own explanations) continued to believe whatever they had initially read. In short, if you give someone half a chance to make up their own reasons to believe something, they'll take you up on the opportunity and start to believe it — even if their original evidence is thoroughly discredited. Rational man, if he (or she) existed, would only believe what is true, invariably moving from true premises to true conclusions. Irrational man, kluged product of evolution that he (or she) is, frequently moves in the opposite direction, starting with a conclusion and seeking reasons to believe it.

Belief, I would suggest, is stitched together out of three fundamental components: a capacity for memory (beliefs would be of no value if they came and went without any long-term hold on the mind), a capacity for inference (deriving new facts from old, as just discussed), and a capacity for, of all things, *perception.*

Superficially, one might think of perception and belief as separate. Perception is what we see and hear, taste, smell, or feel, while belief is what we know or think we know. But in terms of evolutionary history, the two are not as different as they initially appear. The surest path to belief is to see something. When my wife's golden retriever, Ari, wags his tail, I believe him to be happy; mail falls through the slot, and I believe the mail has arrived. Or, as Chico Marx put it, "Who are you gonna believe, me or your own eyes?"

The trouble kicks in when we start to believe things that we don't directly observe. And in the modern world, much of what we believe is not directly or readily observable. Our capacity to acquire new beliefs vicariously — from friends, teachers, or the media, without direct experience — is a key to what allows humans to build cultures and technologies of fabulous complexity. My canine friend Ari learns whatever he learns primarily through trial and error; I learn what I learn mainly through books, magazines, and the Internet. I may cast some skepticism on what I read. (Did journalist-investigator Seymour Hersh really have a well-placed, anonymous source? Did movie reviewer Anthony Lane really even see *Clerks II?*) But largely, for better or worse, I tend to believe what I read, and I learn much of what I know through that medium. Ari (also for better or worse) knows only what he sees, hears, feels, tastes, or smells.

In the early 1990s, the psychologist Daniel Gilbert, now well known for his work on happiness, tested a theory that he traced back to the seventeenth-century philosopher Baruch de Spinoza. Spinoza's idea was that "all information is [initially] accepted during comprehension and . . . false information . . . unaccepted [only later]." As a test of Spinoza's hypothesis, Gilbert presented subjects with true and false propositions — sometimes interrupting them with a brief, distracting tone (which required them to press a button). Just as Spinoza might have predicted, interruptions increased the chance that subjects would believe the false proposition;* other studies showed that people are more likely to accept falsehoods if they are distracted or put under time pressure. The ideas we encounter are, other things being equal, automatically believed — unless and until there is a chance to properly evaluate them.

This difference in order (between hearing, accepting, and evaluating versus hearing, evaluating, and then accepting) might initially seem trivial, but it has serious consequences. Take, for example, a case that was recently described on Ira Glass's weekly radio show *This American Life.* A lifelong political activist who was the leading candidate for chair of New Hampshire's Democratic Party was accused of possessing substantial amounts of child pornography. Even though his accuser, a Republican state representative, offered no proof, the accused was forced to step down, his political career essentially ruined. A two-month investigation ultimately found no evidence, but the damage was done — our legal system may be designed around the principle of "innocent until proven guilty," but our mind is not.

Indeed, as every good lawyer knows intuitively, just *asking* about some possibility can increase the chance that someone will believe it. ("Isn't it true you've been reading pornographic magazines since you were twelve?" "Objection — irrelevant!") Experimental evidence bears this out: merely hearing something in the form of a question —

---

*The converse wasn't true: interrupting people's consideration of *true* propositions didn't lead to increased disbelief precisely because people initially accept that what they hear is true, whether or not they ultimately get a chance to properly evaluate it.

rather than a declarative statement — is often enough to induce belief.

Why do we humans so often accept uncritically what we hear? Because of the way in which belief evolved: from machinery first used in the service of perception. And in perception, a high percentage of what we see is true (or at least it was before the era of television and Photoshop). When we see something, it's usually safe to believe it. The cycle of belief works in the same way — we gather some bit of information, directly, through our senses, or perhaps more often, indirectly through language and communication. Either way, we tend to immediately believe it and only later, if at all, consider its veracity.

The trouble with extending this "Shoot first, ask questions later" approach to belief is that the linguistic world is much less trustworthy than the visual world. If something looks like a duck and quacks like a duck, we are licensed to think it's a duck. But if some guy in a trenchcoat tells us he wants to *sell* us a duck, that's a different story. Especially in this era of blogs, focus groups, and spin doctors, language is not always a reliable source of truth. In an ideal world, the basic logic of perception (*gather information, assume true, then evaluate if there is time*) would be inverted for explicit, linguistically transmitted beliefs; but instead, as is often the case, evolution took the lazy way out, building belief out of a progressive overlay of technologies, consequences be damned. Our tendency to accept what we hear and read with far too little skepticism is but one more consequence.

Yogi Berra once said that 90 percent of the game of baseball was half mental; I say, 90 percent of what we believe is half cooked. Our beliefs are contaminated by the tricks of memory, by emotion, and by the vagaries of a perceptual system that really ought be fully separate — not to mention a logic and inference system that is as yet, in the early twenty-first century, far from fully hatched.

The dictionary defines the act of *believing* both as "accepting something as true" and as "being of the opinion that something exists, especially when there is no absolute proof." Is belief about what we *know* to be true or what we *want* to be true? That it is so often difficult for members of our species to tell the difference is a pointed reminder of our origins.

Evolved of creatures that were often forced to act rather than think, *Homo sapiens* simply never evolved a proper system for keeping track of what we know and how we've come to know it, uncontaminated by what we simply wish were so.

4

CHOICE

People behave sometimes as if they had two selves, one who wants clean lungs and long life and another who adores tobacco, one who yearns to improve himself by reading Adam Smith on self-command (in *The Theory of Moral Sentiments*) and another who would rather watch an old movie on television. The two are in continual contest for control.

— THOMAS SCHELLING

IN THE LATE 1960s and early 1970s, in the midst of the craze for the TV show *Candid Camera* (forerunner of YouTube, reality TV, and shows like *America's Funniest Home Videos*), the psychologist Walter Mischel offered four-year-old preschoolers a choice: a marshmallow now, or two marshmallows if they could wait until he returned. And then, cruelly, he left them alone with nothing more than themselves, the single marshmallow, a hidden camera, and no indication of when he would return. A few of the kids ate the oh-so-tempting marshmallow the minute he left the room. But most kids wanted the bigger bonus and endeavored to wait. So they tried. Hard. But with nothing else to do in the room, the torture was visible. The kids did just about anything they could to distract themselves from the tempting marshmallow that stood before them: they talked to themselves, bounced up and down, covered their eyes, sat on their hands — strategies that more than a few adults might on occasion profitably adopt. Even so, for about half the kids, the 15 or 20 minutes until Mischel returned was just too long to wait.

Giving up after 15 minutes is a choice that could only really make sense under two circumstances: (1) the kids were so hungry that having the marshmallow now could stave off true starvation or (2) their prospects for a long and healthy life were so remote that the 20-minute future versions of themselves, which would get the two marshmallows, simply weren't worth planning for. Barring these rather remote possibilities, the children who gave in were behaving in an entirely irrational fashion.

Toddlers, of course, aren't the only humans who melt in the face of temptation. Teenagers often drive at speeds that would be unsafe even on the autobahn, and people of all ages have been known to engage in unprotected sex with strangers, even when they are aware of the risks. The preschoolers' marshmallows have a counterpart in my raspberry cheesecake, which I know I'll regret later but nevertheless want desperately now. If you ask people whether they'd rather have a certified check for $100 that they can cash now, or a check for twice as much that they can't cash for three years, more than half will take the $100 now. (Curiously — and I will come back to this later — most people's preferences *reverse* when the time horizon is lengthened, preferring $200 in nine years to $100 in six years.) Then there are the daily uncontrollable choices made by alcoholics, drug addicts, and compulsive gamblers. Not to mention the Rhode Island convict who attempted to escape from jail on day 89 of a 90-day prison sentence.

Collectively, the tendencies I just described exemplify what philosophers call "weakness of the will," and they are our first hint that the brain mechanisms that govern our everyday choices might be just as kluge-y as those that govern memory and belief.

Wikipedia defines *Homo economicus,* or Economic man, as the assumption, popular in many economic theories, that man is "a rational and self-interested actor who desires wealth, avoids unnecessary labor, and has the ability to make judgments towards those ends."

At first glance, this assumption seem awfully reasonable. Who among us isn't self-interested? And who wouldn't avoid unnecessary labor, given the chance? (Why clean your apartment unless you know that guests are coming?)

But as the architect Mies van der Rohe famously said, "God is in the details." We are indeed good at dodging unnecessary labor, but true rationality is an awfully high standard, frequently well beyond our grasp. To be truly rational, we would need, at a minimum, to face each decision with clear eyes, uncontaminated by the lust of the moment, prepared to make every decision with appropriately dispassionate views of the relevant costs and benefits. Alas, as we'll see in a moment, the weight of the evidence from psychology and neuroscience suggests otherwise. We *can* be rational on a good day, but much of the time we are not.

Appreciating what we as a species can and can't do well — when we are likely to make sound decisions and when we are likely to make a hash of them — requires moving past the idealization of economic man and into the more sticky territory of human psychology. To see why some of our choices appear perfectly sensible and others perfectly foolish, we need to understand how our capacity for choice *evolved.*

I'll start with good news. On occasion, human choices can be entirely rational. Two professors at NYU, for example, studied what one might think of as the world's simplest touch-screen video game — and found that, within the parameters of that simple task, people were almost as rational (in the sense of maximizing reward relative to risk) as you could possibly imagine. Two targets appear (standing still) on a screen, one green, one red. In this task, you get points if you touch the green circle; you lose a larger number of points if you touch the red one. The challenge comes when the circles overlap, as they often do, and if you touch the intersection between the circles, you get both the reward and the (larger) penalty, thus accruing a net loss. Because people are encouraged to touch the screen quickly, and because nobody's hand-eye coordination is perfect, the optimal thing to do is to point somewhere *other than* the center of the green circle. For example, if the green circle overlaps but is to the right of the red circle, pointing to the center of the green circle is risky business: an effort to point at the exact center of the green circle will sometimes wind up off target, left of center, smack in the point-losing region where the green and red circles overlap. Instead, it makes more sense to point somewhere to *the right of the center of the green circle,* keeping the probability of hitting the green circle high, while minimizing the probability of hitting the red circle. Somehow people figure all this out, though not necessarily in an explicit or conscious fashion. Even more remarkably, they do so in a manner that is almost perfectly calibrated to the *specific accuracy of their own individual system of hand-eye coordination.* Adam Smith couldn't have asked for more.

The bad news is that such exquisite rationality may well be the exception rather than the rule. People are as good as they are at the pointing-at-circles task because it draws on a mental capacity — the ability to reach for things — that is truly ancient. Reaching is close to a reflex, not just for humans, but for every animal that grabs a meal to bring it closer to its mouth; by the time we are adults, our reaching system is so well tuned, we never even think about it. For instance, in a strict technical sense, every time I reach for my cup of tea, I make a set of choices. I decide that I want the tea, that the potential pleasure and the hydration offered by the beverage outweigh the risk of spillage. More than that, and even less consciously, I *decide* at what angle to send my hand. Should I use my left hand (which is closer) or my right hand (which is better coordinated)? Should I grab the cylindrical central portion of the mug (which holds the contents that I really want) or go instead for the handle, a less direct but easier-to-grasp means to the tea that is inside? Our hands and muscles align themselves automatically, my fingers forming a pincer grip, my elbow rotating so that my hand is in perfect position. Reaching, central to life, involves many decisions, and evolution has had a long time to get them just right.

But economics is not supposed to be a theory of how people reach for coffee mugs; it's supposed be a theory of how they spend their money, allocate their time, plan for their retirement, and so forth — it's supposed to be, at least in part, a theory about how people make *conscious* decisions.

And often, the closer we get to conscious decision making, a more recent product of evolution, the worse our decisions become. When the NYU professors reworked their grasping task to make it a more explicit word problem, most subjects' performance fell to pieces. Our more recently evolved deliberative system is, in this particular respect, no match for our ancient system for muscle control. Outside that rarefied domain, there are loads of circumstances in which human performance predictably defies any reasonable notion of rationality.

Suppose, for example, that I give you a choice between participating in two lotteries. In one lottery, you have an 89 percent chance of winning $1 million, a 10 percent chance of winning $5 million, and a 1 percent chance of winning nothing; in the other, you have a 100 percent chance of winning $1 million. Which do you go for? Almost everyone takes the sure thing.

Now suppose instead your choice is slightly more complicated. You can take either an 11 percent chance at $1 million or a 10 percent chance of winning $5 million. Which do you choose? Here, almost everyone goes for the second choice, a 10 percent shot at $5 million.

What would be the rational thing to do? According to the theory of rational choice, you *should* calculate your "expected utility," or expected gain, essentially averaging the amount you would win across all the possible outcomes, weighted by their probability. An 11 percent chance at $1 million works out to an expected gain of $110,000; 10 percent at $5 million works out to an expected gain of $500,000, clearly the better choice. So far, so good. But when you apply the same logic to the first set of choices, you discover that people behave far less rationally. The expected gain in the lottery that is split 89 percent/10 percent/1 percent is $1,390,000 (89 percent of $1 million plus 10 percent of $5 million plus 1 percent of $0), compared to a mere million for the sure thing. Yet nearly everyone goes for the million bucks — leaving close to half a million dollars on the table. Pure insanity from the perspective of "rational choice."

Another experiment offered undergraduates a choice between two raffle tickets, one with 1 chance in 100 to win a $500 voucher toward a trip to Paris, the other, 1 chance in 100 to win a $500 voucher toward college tuition. Most people, in this case, prefer Paris. No big problem there; if Paris is more appealing than the bursar's office, so be it. But when the odds increase from 1 in 100 to 99 out of 100, most people's preferences *reverse;* given the near certainty of winning, most students suddenly go for the tuition voucher rather than the trip — sheer lunacy, if they'd really rather go to Paris.

To take an entirely different sort of illustration, consider the simple question I posed in the opening chapter: would you drive across town to save $25 on a $100 microwave? Most people would say yes, but hardly anybody would drive across town to save the same $25 on a $1,000 television. From the perspective of an economist, this sort of thinking too is irrational. Whether the drive is worth it should depend on just two things: the value of your time and the cost of the gas, nothing else. Either the value of your time and gas is less than $25, in which case you should make the drive, or your time and gas are worth more than $25, in which case you shouldn't make the drive — end of story. Since the labor to drive across town is the same in both cases and the monetary amount is the same, there's no rational reason why the drive would make sense in one instance and not the other.

On the other hand, to anyone who *hasn't* taken a class in economics, saving $25 on $100 seems like a good deal ("I saved 25 percent!"), whereas saving $25 on $1,000 appears to be a stupid waste of time ("You drove all the way across town to get 2.5 percent off? You must have nothing better to do"). In the clear-eyed arithmetic of the economist, a dollar is a dollar is a dollar, but most ordinary people can't help but think about money in a somewhat less rational way: not in absolute terms, but in *relative* terms.

What leads us to think about money in (less rational) relative terms rather than (more rational) absolute terms?

To start with, humans didn't evolve to think about numbers, much less money, at all. Neither money nor numerical systems are omnipresent. Some cultures trade only by means of barter, and some have simple counting systems with only a few numerical terms, such as *one, two, many.* Clearly, both counting systems and money are cultural inventions. On the other hand, all vertebrate animals are built with what some psychologists call an "approximate system" for numbers, such that they can distinguish more from less. And that system in turn has the peculiar property of being "nonlinear": the difference between 1 and 2 subjectively seems greater than the difference between 101 and 102. Much of the brain is built on this principle, known as Weber's law. Thus, a 150-watt light bulb seems only a bit brighter than a 100-watt bulb, whereas a 100-watt bulb seems much brighter than a 50-watt bulb.

In some domains, following Weber's law makes a certain amount of sense: a storehouse of an extra 2 kilos of wheat relative to a baseline of 100 kilos isn't going to matter if everything after the first kilos ultimately spoils; what really matters is the difference between starving and not starving. Of course, money doesn't rot (except in times of hyperinflation), but our brain didn't evolve to cope with money; it evolved to cope with *food.*

And so even today, there's some remarkable crosstalk between the two. People are less likely to donate money to charities, for example, if they are hungry than if they are full; meanwhile, experimental subjects (excluding those who were dieting) who are put in a state of "high desire for money" eat more M&Ms during a taste test than do people who are in a state of "low desire for money."* To the degree that our understanding of money is kluged onto our understanding of food, the fact that we think about money in relative terms may be little more than one more accident of our cognitive history.

"Christmas Clubs," accounts into which people put away small amounts of money all year, with the goal of having enough money for Christmas shopping at the end of the year, provide another case in point. Although the goal is admirable, the behavior is (at least from the perspective of classical economics) irrational: Christmas Club accounts generally have low balances, so they tend to earn less interest than if the money were pooled with a person's other funds. And in any event, that money, sitting idle, might be better spent paying down high-interest credit card debt. Yet people do this sort of thing all the time, establishing real or imaginary accounts for different purposes, as if the money weren't all theirs.

Christmas Clubs and the like persist not because they are fiscally rational but because they are an accommodation to the idiosyncratic structure of our evolved brain: they provide a way of coping with the weakness of the will. If our self-control were better, we wouldn't need such accommodations. We would save money all year long in a unified account that receives the maximum rate of return, and draw on it as needed; only because the temptation of the present so

---

*How do you get people to lust after money? Let them imagine they'll win a significant amount of money in the lottery, then ask them to think about how they would spend it. The more money they are asked to envision, the more lust is engendered. In the particular experiment I describe, people in the "high desire for money" condition spent a few minutes thinking about how they'd spend a £25,000 prize, while people in the "low desire for money" condition spent a few minutes contemplating how they'd spend a £25 prize. A remarkable indicator of the influence of induced money lust came from a follow-up question: people were asked to estimate the size of coins. The bigger the lust, the larger their estimate.

often outweighs the abstract reality of the future do we fail to do so such a simple, fiscally sound thing. (The temptation of the present also tends to leave our future selves high and dry. According one estimate, nearly two thirds of all Americans save too little for retirement.)

Rationality also takes a hit when we think about so-called sunk costs. Suppose, for instance, that you decide to see a play and plop down $20 for a ticket — only to find, as you enter the theater, that you've lost the ticket. Suppose, further, that you were to be seated in general admission (that is, you have no specific assigned seat), and there's no way to get the ticket back. Would you buy another ticket? Laboratory data show that half the people say yes, while the other half give up and go home, a 50-50 split; fair enough. But compare that scenario with one that is only slightly different. Say you've lost cash rather than a prepurchased ticket. ("Imagine that you have decided to see a play, and the admission is $20 per ticket. As you enter the theater, ready to purchase one, you discover that you have lost a $20 bill. Would you still pay $20 for a ticket for the play?") In this case, a whopping 88 percent of those tested say yes — even though the extra out-of-pocket expense, $20, is identical in the two scenarios.

Here's an even more telling example. Suppose you spend $100 for a ticket to a weekend ski trip to Michigan. Several weeks later you buy a $50 ticket for another weekend ski trip, this time to Wisconsin, which (despite being cheaper) you actually think you'll enjoy more. Then, just as you are putting your newly purchased Wisconsin ski-trip ticket in your wallet, you realize you've goofed: the two trips take place on the same weekend! And it's too late to sell either one. Which trip do you go on? More than half of test subjects said they would choose (more expensive) Michigan — even though they knew they would enjoy the Wisconsin option more. With the money for both trips already spent (and unrecoverable), this choice makes no sense; a person would get more utility (pleasure) out of the trip to Wisconsin for no further expense, but the human fear of "waste" convinces him or her to select the less pleasurable trip.* On a global scale, the same kind of dubious reasoning can have massive consequences. Even presidents have been known to stick to policies long after it's evident to everyone that those policies just aren't working.

Economists tell us that we should assess the value of a thing according to its expected utility, or how much pleasure it will bring,† buying only if the utility exceeds the asking price. But here again, human behavior diverges from economic rationality. If the first principle of how people determine value is that they do so in relative terms (as we saw in the previous section), the second is that people often have only the faintest idea of what something is truly worth.

Instead, we often rely on secondary criteria, such as how good a deal we think we're getting. Consider, for example, the question

---

*As a final illustration, borrowed from maverick economist Richard Thaler, imagine buying an expensive pair of shoes. You like them in the store, wear them a couple of times, and then, sadly, discover that they don't actually fit properly. What happens next? Based on his data, Thaler predicts the following: (1) The more you paid for the shoes, the more times you will try to wear them. (2) Eventually you stop wearing the shoes, but you do not throw them away. The more you paid for the shoes, the longer they will sit in the back of your closet before you dispose of them. (3) At some point, you throw the shoes away, regardless of what they cost, the payment having been fully "depreciated." As Thaler notes, wearing the shoes a few more times might be rational, but holding on to them makes little sense. (My wife, however, notes that your feet could shrink. Or, she adds brightly, "You never know, you might get some sort of foot surgery." Never give up on a nice pair of shoes!)

†This is not to say that prices need to be fixed, as they are on *The Price Is Right*. On this long-running show, the eternally young Bob Barker would, at the end of every segment (until he finally retired in June 2007), intone that "the actual retail value" of some product was such and such: $242 for this watch, $32,733 for this car. But in real life prices don't work that way. Some prices really are fixed, but most vary to some degree or other. Even as a child I found the whole concept puzzling. How could Barker tell us that a Hershey bar cost 30¢ — didn't it matter which store he got the candy from? After all, as I well knew, the neighborhood convenience store charged more than the supermarket. (No economist would object to this particular discrepancy; if you need milk at 2 A.M., and the convenience store is the only place open, it makes sense to pay a premium, and in this regard humans are perfectly rational.)

posed in Bob Merrill's classic sing-along: "How much is that doggie in the window?" How much *is* a well-bred doggie worth? Is a golden retriever worth a hundred times the price of a movie? A thousand times? Twice the value of a trip to Peru? A tenth of the price of a BMW convertible? Only an economist would ask.

But what people actually do is no less weird, often giving more attention to the salesperson's jabber than the dog in question. If the breeder quotes a price of $600 and the customer haggles her down to $500, the customer buys the dog and counts himself lucky. If the salesperson starts at $500 and doesn't budge, the customer may walk out in a huff. And, most likely, that customer is a fool. Assuming the dog is healthy, the $500 probably would have been well spent.*

To take another example, suppose you find yourself on a beach, on a hot day, with nothing to drink — but a *strong* desire for a nice cold beer. Suppose, furthermore, that a friend of yours kindly offers to get you a beer, provided that you front him the money. His only request is that you tell him — in advance — the most you are willing to pay; your friend doesn't want to have the responsibility of deciding for you. People often set their limit according to where the beer is going to be purchased; you might go as high as $6 if the beer is to be purchased at a resort, but no more than $4 if the friend were going to a bodega at the end of the beach. From an economist's perspective, that's just loopy: the true measure should be "How much pleasure would that beer bring me?" and not "Is the shop/resort charging a price that is fair relative to other similar establishments?" Six dollars is $6, and if the beer would bring $10 of pleasure, $6 is a bargain, even if one spends it at the world's most expensive bodega. In the dry language of one economist, "The consumption experience is the same."

---

*I'm neither a dog owner nor a "dog person," but if my wife's experience is at all typical, a fluffy golden may be among the best things money can buy. She's had hers for a dozen years, and he still brings her pleasure every day, far more than I can say for any of the myriad electronic gadgets I've ever acquired.

The psychologist Robert Cialdini tells a story of a shopkeeper friend of his who was having trouble moving a certain set of necklaces. About to go away for vacation, this shopkeeper left a note for her employees, intending to tell them to cut the price in half. Her employees, who apparently had trouble reading the note, instead *doubled* the price. If the necklaces didn't budge at $100, you'd scarcely expect them to sell at $200. But that's exactly what happened; by the time the shopkeeper had returned, the whole inventory was gone. Customers were more likely to buy a particular necklace if it had a high price than if it had a low price — apparently because they were using list price (rather than intrinsic worth) as a proxy for value. From the perspective of economics, this is madness.

What's going on here? These last few examples should remind you of something we saw in the previous chapter: anchoring. When the value we set depends on irrelevancies like a shopkeeper's starting price as much as it does on an object's intrinsic value, anchoring has clearly cluttered our head.

Anchoring is such a basic part of human cognition that it extends not just to how we value puppies or material goods, but even to intangibles like life itself. One recent study, for example, asked people how much they would pay for safety improvements that would reduce the annual risk of automobile fatalities. Interviewers would start by asking interviewees whether they would be willing to pay some fairly low price, either £25 or £75. Perhaps because nobody wished to appear to be a selfish lout, answers were always in the affirmative. The fun came after: the experimenter would just keep pushing and pushing until he (or she) found a given subject's upper limit. When the experimenter started with £25 per year, subjects could be driven up to, on average, £149. In contrast, when the experimenter started at £75 per year, subjects tended to go up almost 40 percent higher to an average maximum of £232.

Indeed, virtually *every* choice that we make, economic or not, is, in some way or another, influenced by how the problem is posed. Consider, for example, the following scenario:

> Imagine that the nation is preparing for the outbreak of an unusual disease, which is expected to kill 600 people. Two alternative programs to combat the disease have been proposed. Assume that the exact scientific estimates of the consequences of the programs are as follows:
>
> If Program A is adopted, 200 people will be saved.
>
> If Program B is adopted, there is a one-third probability that 600 people will be saved and a two-thirds probability that no people will be saved.

Most people would choose Program A, not wanting to put all the lives at risk. But people's preferences flip if the same choices are instead posed this way:

> If Program A is adopted, 400 people will die.
>
> If Program B is adopted, there is a one-third probability that nobody will die and a two-thirds probability that 600 people will die.

"Saving 200 lives" for certain (out of 600) somehow seems like a good idea, whereas letting 400 die (out of the same 600) seems bad — even though they represent exactly the same outcome. Only the wording of the question, what psychologists call *framing*, has been changed.

Politicians and advertisers take advantage of our susceptibility to framing all the time. A death tax sounds far more ominous than an inheritance tax, and a community that is described as having a crime rate of 3.7 percent is likely to get more resources than one that is described as 96.3 percent crime free.

Framing has the power that it does because choice, like belief, is inevitably mediated by memory. And, as we have already seen, the memory that evolution equipped us with is inherently and inevitably shaped by momentary contextual details. Change the context (here, the actual words used), and you often change the choice. "Death tax" summons thoughts of death, a fate that we all fear, whereas "inheritance tax" may make us think only of the truly wealthy, suggesting a tax scarcely relevant to the average taxpayer. "Crime rates" makes us think of crime; "crime-free rates" triggers thoughts of safety. What we think of — what we summon into memory as we come to a decision — often makes all the difference.

Indeed, the whole field of advertising is based on that premise: if a product brings pleasant associations to mind, no matter how irrelevant, you're more likely to buy it.*

One Chicago law firm recently put the power of memory and suggestion to the ultimate test, flogging not potato chips or beer but the dissolution of marriage. Their tool? A 48-foot-wide billboard made of three panels — the torso of an exceptionally attractive woman, breasts all but bursting out of her lacy black bra; the torso of a man no less handsome, shirtless, with his well-oiled muscles bulging; and, just above the law firm's name and contact information, a slogan containing just five words: LIFE'S SHORT — GET A DIVORCE.

In a species less driven by contextual memory and spontaneous priming, I doubt that sign would have any impact. But in a species like ours, there's reason to worry. To seek a divorce is, of course, to make one of the most difficult choices a human being can make. One must weigh hopes for the future against fears of loneliness, regret, financial implications, and (especially) concerns about children. Few

---

*The future of advertising on the Internet is no doubt going to revolve around *personalized* framing. For example, some people tend to focus on achieving ideals (what is known in the literature as having a "promotion focus") while others tend toward a "prevention focus," aimed at avoiding failure. People with promotion focus may be more responsive to appeals pitched in terms of a given product's advantages, while people with a prevention focus may be more responsive to pitches emphasizing the cost of making do *without* the product.

people make such decisions lightly. In a rational world, a titillating billboard wouldn't make a dime's difference. In the real world of flesh-and-blood human beings governed by klugey brains, people who weren't otherwise thinking of divorce might well be induced to start. What's more, the billboard might frame *how* people think about divorce, leading them to evaluate their marriage not in terms of companionship, family, and financial security, but whether it includes enough ripped bodices and steamy sexual encounters.

If this seems speculative, that's because the law firm took the sign down, under pressure, after just a couple of weeks, so there's no direct evidence. But a growing literature of real-world marketing studies backs me up. One study, for example, asked people how likely they were to buy a car in the next six months. People who were asked whether they'd buy a car were almost twice as likely to actually do so than those who weren't asked. (Small wonder that many car dealers ask not *whether* you are going to buy a car but *when*.) The parallel to a lawyer's leading question is exact, the mechanism the same. Just as context influences belief by jostling the current contents of our thoughts, it also affects choice.

The cluster of phenomena I've just discussed — framing, anchoring, susceptibility to advertising, and the like — is only part of the puzzle; our choices are also contaminated by memories retrieved from *within*. Consider, for example, a study that examined how office workers, some feeling hungry, some not, would select which snack they'd like to have a week hence, in the late afternoon. Seventy-two percent of those who were hungry at the time of the decision (several days before they would be having the snack in question) chose unhealthful snacks, like potato chips or candy bars. Among the people who weren't feeling hungry, only 42 percent chose the same unhealthful snacks; most instead committed themselves to apples and bananas. Everybody knows an apple is a better choice (consistent with our long-term goal of staying healthy), but when we feel hungry, memories of the joys of salt and refined sugar win out.

All of this is, of course, a function of evolution. Rationality, pretty much by definition, demands a thorough and judicious balancing of evidence, but the circuitry of mammalian memory simply isn't attuned to that purpose. The speed and context-sensitivity of memory no doubt helped our ancestors, who had to make snap decisions in a challenging environment. But in modern times, this former asset has become a liability. When context tells us one thing, but rationality another, rationality often loses.

Evolutionary inertia made a third significant contribution to the occasional irrationality of modern humans by calibrating us to expect a degree of uncertainty that is largely (and mercifully) absent in contemporary life. Until very recently, our ancestors could not count on the success of next year's harvest, and a bird in hand was certainly better than two, or even three, in the bush. Absent refrigerators, preservatives, and grocery stores, mere survival was far less assured than it is today — in the immortal words of Thomas Hobbes, life was "nasty, brutish, and short."

As a result, over hundreds of millions of years, evolution selected strongly for creatures that lived largely in the moment. In every species that's ever been studied, animals tend to follow what is known as a "hyperbolic discounting curve" — fancy words for the fact that organisms tend to value the present far more than the future. And the closer temptation is, the harder it is to resist. For example, at a remove of 10 seconds, a pigeon can recognize (so to speak) that it's worth waiting 14 seconds to get four ounces of food rather than a single ounce in 10 seconds — but if you wait 9 seconds and let the pigeon change its choice at the last moment, it will. At the remove of just 1 second, the desire for food *now* overwhelms the desire for more food later; the pigeon refuses to wait an extra 4 seconds, like a hungry human noshing on chips while he waits for dinner to arrive.

Life is generally much more stable for humans than for the average pigeon, and human frontal lobes much larger, but still we humans can't get over the ancestral tendency to live in the moment. When we are hungry, we gobble French fries as if driven to lard up on carbs and fat now, since we might not find any next week. Obesity is chronic not just because we routinely underexercise, but also because our brain hasn't caught up with the relative cushiness of modern life.* We continue to discount the future enormously, even as we live in a world of all-night grocery stores and 24/7 pizza delivery.

Future discounting extends well beyond food. It affects how people spend money, why they fail to save enough for retirement, and why they so frequently rack up enormous credit card debt. One dollar now, for example, simply seems more valuable than $1.20 a year hence, and nobody seems to think much about how quickly compound interest rises, precisely because the subjective future is just so far away — or so we are evolved to believe. To a mind not evolved to think about money, let alone the future, credit cards are almost as serious a problem as crack. (Fewer than 1 in 50 Americans uses crack regularly, but nearly half carry regular credit card debt, almost 10 percent owing over $10,000.)

Our extreme favoritism toward the present at the expense of the future would make sense if our life span were vastly shorter, or if the world were much less predictable (as was the case for our ancestors), but in countries where bank accounts are federally insured and grocery stores reliably restocked, the premium we place on the present is often seriously counterproductive.

The more we discount the future, the more we succumb to short-term temptations like drugs, alcohol, and overeating. As one researcher, Howard Rachlin, sums it up,

> in general, living a healthy life for a period of ten years, say, is intrinsically satisfying . . . Over a ten-year period, virtually all

---

*Ironically, our ability to moderate temptation increases with age, even as our life expectancy goes down. Children, who are the most likely to live into the future, are the least likely to be patient enough to wait for it.

would prefer living a healthy life to being a couch potato. Yet we also (more or less) prefer to drink *this drink* than not to drink it, to eat *this chocolate sundae* than to forgo it, to smoke *this cigarette* than not smoke it, to watch *this TV program* than spend a half-hour exercising . . . [emphasis added]

I don't think it's exaggerating to say that this tension between the short term and the long term defines much of contemporary Western life: the choice between going to the gym now and staying home to watch a movie, the joy of the French fries now versus the pain of winding up later with a belly the size of Buddha's.

But the notion that we are shortsighted in our choices actually explains only *half* of this modern bourgeois conflict. The other half of the story is that we humans are the only species smart enough to appreciate the fact that there is another option. When the pigeon goes for the one ounce now, I'm not sure it feels any remorse at what has been lost. I, on the other hand, have shown myself perfectly capable of downing an entire bag of the ironically named Smartfood popcorn, even as I recognize that in a few hours I will regret it.

And that too is a sure sign of a kluge: when I can do something stupid *even as I know at the time* that it's stupid, it seems clear that my brain is a patchwork of multiple systems working in conflict. Evolution built the ancestral reflexive system first and evolved systems for rational deliberation second — fine in itself. But any good engineer would have put some thought into integrating the two, perhaps largely or entirely turning over choices to the more judicious human forebrain (except possibly during time-limited emergencies, where we have to act without the benefit of reflection). Instead, our ancestral system seems to be the default option, our first recourse just about all the time, whether we need it or not. We eschew our deliberative system not just during a time crunch, but also when we are tired, distracted, or just plain lazy; using the deliberative system seems to require an act of will. Why? Perhaps it's simply because the older system came first, and — in systems built through the progressive overlay of technology — what comes first tends to remain intact. And no matter how shortsighted it is, our deliberative system (if it manages to get involved at all) inevitably winds up contaminated. Small wonder that future discounting is such a hard habit to shake.

Choice slips a final cog when it comes to the tension between logic and emotion. The temptation of the immediate present is but one example; many alcoholics know that continued drink will bring them to ruin, but the anticipated pleasure in a drink at a given moment is often enough to overwhelm sensible choice. Emotion one, logic zero.

Perhaps it is only a myth that Menelaus declared war on the Trojans after Paris abducted the woman Menelaus loved, but there can be little doubt that some of the most significant decisions in history have been made for reasons more emotional than rational. This may well, for example, have been the case in the 2003 invasion of Iraq; only a few months earlier, President Bush was quoted as saying, in reference to Saddam Hussein, "After all, this is the guy who tried to kill my dad." Emotion almost certainly plays a role when certain individuals decide to murder their spouse, especially one caught in flagrante delicto. Positive emotions, of course, influence many decisions too — the houses people buy, the partners they marry, the sometimes dubious individuals with whom they have short-term flings. As my father likes to say, "All sales" — and indeed all choices — "are emotional."

From the perspective developed in this book, what is klugey is not so much the fact that people sometimes rely on emotions but rather the way those emotions *interact with* the deliberative system. This is true not just in the obvious scenarios I mentioned — those involving jealousy, love, vengeance, and so forth — but even in cases that don't appear to engage our emotions at all. Consider, for example, a study that asked people how much they would contribute toward various environmental programs, such as saving dolphins or providing free medical checkups to farm workers in order to reduce the incidence of skin cancer. When asked which effort they thought was more important, most people point to the farm workers (perhaps because they valued human lives more than those of dolphins). But when researchers asked people how much *money* they would donate to each cause, dolphins and farm workers, they gave more to the cuddly dolphins. Either choice on its own might make sense, but making the two together is as inconsistent as you can get. Why would someone spend more money on dolphins if that person thinks that human lives are more important? It's one thing for our deliberative system to be out of sync with the ancestral system, another for the two to flip-flop arbitrarily in their bid for control.

In another recent study, people were shown a face — happy, sad, or neutral — for about a sixtieth of a second — and then were asked to drink a "novel lemon-lime beverage." People drank more lemon-lime after seeing happy faces than after viewing sad ones, and they were willing to *pay twice as much* for the privilege. All this presumably shows that the process of priming affects our choices just as much as our beliefs: a happy face primes us to approach the drink as if it were pleasant, and a sad face primes us to avoid the drink (as if it were unpleasant). Is it any wonder that advertisers almost always present us with what the rock band REM once called "shiny, happy people"?

An even more disquieting study asked a group of subjects to play a game known as "prisoner's dilemma," which requires pairs of people to choose to either cooperate with each other or "defect" (act uncooperatively). You get the bigger payoff if you and the other person both cooperate (say, $10), an intermediate reward (say, $3) if you defect and your opponent cooperates, and no reward if you both defect. The general procedure is a staple in psychology research; the catch in this particular study was that before people began to play the game, they sat in a waiting room where an ostensibly unrelated news broadcast was playing in the background. Some subjects heard prosocial news (about a clergyman donating a kidney to a needy patient); others, by contrast, heard a broadcast about a clergyman committing murder. What happened? You guessed it: people who heard about the good clergyman were a lot more cooperative than those who heard about the bad clergyman.

In all these studies, emotions of one sort or another prime memories, and those memories in turn shape choice. A different sort of illustration comes from what economist George Loewenstein calls "the attraction of the visceral." It's one thing to turn down chocolate cheesecake in the abstract, another when the waiter brings in the dessert cart. College students who are asked whether they'd risk wasting 30 minutes in exchange for a chance to win all the freshly baked chocolate chip cookies they could eat are more likely to say yes if they actually *see* (and smell) the cookies than if they are merely told about them.

Hunger, however, is nothing compared to lust. A follow-up study exposed young men to either a written or a (more visceral) filmed scenario depicting a couple who had met earlier in the evening and are now discussing the possibility of (imminently) having sex. Both are in favor, but neither party has a condom, and there is no store nearby. The woman reports that she is taking a contraceptive pill and is disease-free; she leaves it up to the man to decide whether to proceed, unprotected. Subjects were then asked to rate their own probability of having unprotected sex if they were in the male character's shoes. Guess which group of men — readers or video watchers — was more likely to throw caution to the wind? (Undergraduate men are also apparently able to persuade themselves that their risk of contracting a sexually transmitted disease goes down precisely as the attractiveness of their potential partner goes up.) The notion that men might think with organs below the brain is not new, but the experimental evidence highlights rather vividly the degree to which our choices don't necessarily follow from purely "rational" considerations. Hunger, lust, happiness, and sadness are all factors that most of us would say shouldn't enter into rational thought. Yet evolution's progressive overlay of technology has guaranteed that each wields an influence, even when we insist otherwise.

The clumsiness of our decision-making ability becomes especially clear when we consider moral choices. Suppose, for example, that a runaway trolley is about to run over and kill five people. You (and you alone) are in a position such that you can hit a switch to divert the trolley onto a different set of tracks, where it would kill only one person instead of five. Do you hit the switch?

Now, suppose instead that you are on a footbridge, standing above the track that bears the runaway trolley. This time, saving the five people would require you to push a rather large person (considerably bigger than you, so don't bother to volunteer yourself) off the footbridge and into the oncoming trolley. The large person in question would, should you toss him over, die, allowing the other five to survive. Would *that* be okay? Although most people answer yes to the scenario involving the switch, most people say no to pushing someone off the footbridge — even though in both cases five lives are saved at the cost of one.

Why the difference? Nobody knows for sure, but part of the answer seems to be that there is something more visceral about the second scenario; it's one thing to flip a switch, which is inanimate and somewhat removed from the actual collision, and another to forcibly send someone to his death.

One historical example of how visceral feelings affect moral choice is the unofficial truce called by British and German soldiers during Christmas 1914, early in World War I. The original intention was to resume battle afterward, but the soldiers got to know one another during the truce; some even shared a Christmas meal. In so doing, they shifted from conceptualizing one another as enemies to seeing each other as flesh-and-blood individuals. The consequence was that after the Christmas truce, the soldiers were no longer able to kill one another. As the former president Jimmy Carter put it in his Nobel Peace Prize lecture (2002), "In order for us human beings to commit ourselves personally to the inhumanity of war, we find it necessary first to dehumanize our opponents."

Both the trolley problem and the Christmas truce remind us that though our moral choices may *seem* to be the product of a single process of deliberative reasoning, our gut, in the end, often also plays a huge role, whether we are speaking of something mundane, like a new car, or making decisions with lives at stake.

The trolley scenarios illustrate the split by showing how we can get two different answers to essentially the same question, depending on which system we tap into. The psychologist Jonathan Haidt has tried to go a step further, arguing that we can have strong moral intuitions even when we can't back them up with explicit reasons. Consider, for example, the following scenario:

> Julie and Mark are brother and sister. They are traveling together in France on summer vacation from college. One night they are staying alone in a cabin near the beach. They decide that it would be interesting and fun if they tried making love. At the very least it would be a new experience for each of them. Julie was already taking birth control pills, but Mark uses a condom too, just to be safe. They both enjoy making love, but they decide not to do it again. They keep that night as a special secret, which makes them feel even closer to each other. What do you think about that? Was it okay for them to make love?

Every time I read this passage, I get the creeps. But *why* exactly is it wrong? As Haidt describes it,

> most people who hear the above story immediately say that it was wrong for the siblings to make love, and they then begin searching for reasons. They point out the dangers of inbreeding, only to remember that Julie and Mark used two forms of birth

control. They argue that Julie and Mark will be hurt, perhaps emotionally, even though the story makes it clear that no harm befell them. Eventually, many people say something like "I don't know, I can't explain it, I just know it's wrong."

Haidt calls this phenomenon — where we feel certain that something is wrong but are at a complete loss to explain why — "moral dumbfounding." I call it an illustration of how the emotional and the judicious can easily decouple. What makes moral dumbfounding possible is the split between our ancestral system — which looks at an overall picture without being analytical about the details — and a judicious system, which can parse things piece by piece. As is so often the case, where there is conflict, the ancestral system wins: even though we know we can't give a good reason, our emotional queasiness lingers.

When you look inside the skull, using neuroimaging, you find further evidence that our moral judgments derive from two distinct sources: people's choices on moral dilemmas correlate with how they use their brains. In experimental trials like those mentioned earlier, the subjects who chose to save five lives at the expense of one tended to rely on the regions of the brain known as the dorsolateral prefrontal cortex and the posterior parietal cortex, which are known to be important for deliberative reasoning. On the other hand, people who decided in favor of the single individual at the cost of five tended to rely more on regions of the limbic cortex, which are more closely tied to emotion.*

What makes the human mind a kluge is not the fact that we have two systems per se but the way in which the two systems *interact*. In principle, a deliberative reasoning system should be, well, deliberate: above the fray and unbiased by the considerations of the emotional

---

*Fans of the history of neuroscience will recognize this as the brain region that was skewered in the brain of one Phineas Gage, injured on September 13, 1848.

system. A sensibly designed deliberative-reasoning machine would systematically search its memory for relevant data, pro and con, so that it could make systematic decisions. It would be attuned as much to disconfirmation as confirmation and utterly immune to patently irrelevant information (such as the opening bid of a salesperson whose interests are necessarily different from your own). This system would also be empowered to well and truly stifle violations of its master plan. ("I'm on a diet. No chocolate cake. Period.") What we have instead falls between two systems — an ancestral, reflexive system that is only partly responsive to the overall goals of the organism, and a deliberative system (built from inappropriate old parts, such as contextual memory) that can act in genuinely independent fashion only with great difficulty.

Does this mean that our conscious, deliberate choices are always the best ones? Not at all. As Daniel Kahneman has observed, the reflexive system is better at what *it* does than the deliberative system is at deliberating. The ancestral system, for example, is exquisitely sensitive to statistical fluctuations — its bread and butter, shaped over eons, is to track the probabilities of finding food and predators in particular locations. And while our deliberative system *can* be deliberate, it takes a great deal of effort to get it to function in genuinely fair and balanced ways. (Of course, this is no surprise if you consider that the ancestral system has been shaped for hundreds of millions of years, but deliberative reasoning is still a bit of a newfangled invention.)

So, inevitably, there are decisions for which the ancestral system is better suited; in some circumstances it offers the only real option. For instance, when you have to make a split-second decision — whether to brake your car or swerve into the next lane — the deliberative system is just too slow. Similarly, where we have many different variables to consider, the unconscious mind — given suitable time — can sometimes outperform the conscious deliberative mind; if your problem requires a spreadsheet, there's a chance that the ancestral, statistically inclined mind might be just the ticket. As Malcolm Gladwell said in his recent book *Blink*, "Decisions made very quickly can be every bit as good as decisions made consciously and deliberately."

Still, we shouldn't blindly trust our instincts. When people make effective snap decisions, it's usually because they have ample experience with similar problems. Most of Gladwell's examples, like that of an art curator who instantly recognizes a forgery, come from experts, not amateurs. As the Dutch psychologist Ap Dijksterhuis, one of the world's leading researchers on intuition, noted, our best intuitions are those that are the result of *thorough* unconscious thought, honed by years of experience. Effective snap decisions (Gladwell's "blinks") often represent the icing on a cake that has been baking for a very long time. Especially when we face problems that differ significantly from those that we've faced before, deliberative reasoning can be our first and best hope.

It would be foolish to routinely surrender our considered judgment to our unconscious, reflexive system, vulnerable and biased as it often is. But it would be just as silly to abandon the ancestral reflexive system altogether: it's not entirely irrational, just less reasoned. In the final analysis, evolution has left us with two systems, each with different capabilities: a reflexive system that excels in handling the routine and a deliberative system that can help us think outside the box.

Wisdom will come ultimately from recognizing and harmonizing the strengths and weaknesses of the two, discerning the situations in which our decisions are likely to be biased, and devising strategies to overcome those biases.

5
LANGUAGE

One morning I shot an elephant in my pajamas.
How he got into my pajamas I'll never know.

— GROUCHO MARX

SHE SELLS SEASHELLS by the seashore. A pleasant peasant pheasant plucker plucks a pleasant pheasant. These are words that twist the tongue.

Human language may seem majestic, from the perspective of a vervet monkey confined to a vocabulary of three words (roughly, *eagle, snake,* and *leopard*). But in reality, language is filled with foibles, imperfections, and idiosyncrasies, from the way we pronounce words to the ways we put together sentences. We start, we stop, we stutter, we use *like* as a punctuation marker; we swap our consonants like the good Reverend William Archibald Spooner (1844–1930), who turned Shakespeare's *one fell swoop* into *one swell foop*. (*A real smart feller* becomes a *real . . .* well, you get the idea.) We may say *bridge of the neck* when we really mean *bridge of the nose;* we may mishear *All of the members of the group grew up in Philadelphia* as *All of the members of the group threw up in Philadelphia.* Mistakes like these* are a tic of the human mind.

The challenge, for the cognitive scientist, is to figure out which idiosyncrasies are really important. Most are mere trivia, amusing but not reflective of the deep structures of the mind. The word *driveway,*

---

*If to err is human, to write it down is divine. This chapter is written in memory of the late Vicki Fromkin, an early pioneer in linguistics who was the first to systematically collect and study human speech errors. You can read more about her at http://www.linguistics.ucla.edu/people/fromkin/fromkin.htm.

for example, used to refer to driving on a private road that went from a main road to a house. In truth, we still drive on (or at least into) driveways, but we scarcely notice the *driving* part, since the drive is short; the word's meaning shifted when real estate boomed and our ideas of landscaping changed. (The *park* in *parkways* had nothing to do with parking, but rather with roads that ran along or through parks, woodsy green places that have given way to suburbs and the automobile.) Yet facts like these reveal nothing deep about the mind because other languages are free to do things more systematically, so that cars would park, for example, in a *Parkplatz.*

Likewise it is amusing but not deeply significant to note that we "relieve" ourselves in water closets and bathrooms, even though our W.C.'s are bigger than closets and our bathrooms have no baths. (For that matter, public restrooms may be public, and may be rooms, but I've never seen anyone rest in one.) But our reluctance to say where we plan to go when we "have to go" isn't really a flaw in language; it's just a circumlocution, a way of talking around the details in order to be polite.

Some of the most interesting linguistic quirks, however, run deeper, reflecting not just the historical accidents of particular languages, but fundamental truths about those creatures that produce language — namely, us.

Consider, for instance, the fact that *all* languages are rife with ambiguity, not just the sort we use deliberately ("I can't recommend this person enough") or that foreigners produce by accident (like the hotel that advised its patrons to "take advantage of the chambermaid"), but the sort that ordinary people produce quite by accident, sometimes with disastrous consequences. One such case occurred in 1982, when a pilot's ambiguous reply to a question about his position ("at takeoff") led to a plane crash that killed 583 people; the pilot in question said "Ready for takeoff," but air traffic control interpreted this as meaning "in the process of taking off."

To be perfect, a language would presumably have to be unambiguous (except perhaps where deliberately intended to be ambiguous), systematic (rather than idiosyncratic), stable (so that, say, grandparents could communicate with their grandchildren), nonredundant (so as not to waste time or energy), and capable of expressing any and all of our thoughts.* Every instance of a given speech sound would invariably be pronounced in a constant way, each sentence as clean as a mathematical formula. In the words of one of the leading philosophers of the twentieth century, Bertrand Russell,

> in a logically perfect language, there will be one word and no more for every simple object, and everything that is not simple will be expressed by a combination of words, by a combination derived, of course, from the words for the simple things that enter in, one word for each simple component. A language of that sort will be completely analytic, and will show at a glance the logical structure of the facts asserted or denied.

Every human language falls short of this sort of perfection. Russell was probably wrong in his first point — it's actually quite handy (logical, even) for a language to allow for the household pet to be referred to as *Fido, a dog, a poodle, a mammal,* and *an animal* — but right in thinking that in an ideal language, words would be systematically related in meaning and in sound. But this is distinctly not the case. The words *jaguar, panther, ocelot,* and *puma,* for example, sound totally different, yet all refer to felines, while hardly any of the words that sound like *cat* — *cattle, catapult, catastrophe* — have any connection to cats.

Meanwhile, in some cases language seems redundant (*couch* and *sofa* mean just about the same thing), and in others, incomplete (for example, no language can truly do justice to the subtleties of what we can smell). Other thoughts that seem perfectly coherent can be surprisingly difficult to express; the sentence *Whom do you think that*

---

*Forgive me if I leave poetry out of this. Miscommunication can be a source of mirth, and ambiguity may enrich mysticism and literature. But in both cases, it's likely that we're making the best of an imperfection, not exploiting traits specifically shaped by their adaptive value.

John left? (where the answer is, say, Mary, his first wife) is grammatical, but the ostensibly similar *Whom do you think that left Mary?* (where the answer would be John) is not. (A number of linguists have tried to explain this phenomenon, but it's hard to understand why this asymmetry should exist at all; there's no real analogy in mathematics or computer languages.)

Ambiguity, meanwhile, seems to be the rule rather than the exception. A *run* can mean anything from a jog to a tear in a stocking to scoring a point in baseball, a *hit* anything from a smack to a best-selling tune. When I say "I'll give you a ring tomorrow," am I promising a gift of jewelry or just a phone call? Even little words can be ambiguous; as Bill Clinton famously said, "It all depends on what the meaning of the word '*is*' is." Meanwhile, even when the individual words are clear, sentences as a whole may not be: does *Put the book on the towel on the table* mean that there is a book on the towel that ought to be on the table or that a book, which ought to be on a towel, is already on the table?

Even in languages like Latin, which might — for all its cases and word endings — seem more systematic, ambiguities still crop up. For instance, because the subject of a verb can be left out, the third-person singular verb *Amat* can stand on its own as a complete sentence — but it might mean "He loves," "She loves," or "It loves." As the fourth-century philosopher Augustine, author of one of the first essays on the topic of ambiguity, put it, in an essay written in the allegedly precise language of Latin, the "perplexity of ambiguity grows like wild flowers into infinity."

And language falls short on our other criteria too. Take redundancy. From the perspective of maximizing communication relative to effort, it would make little sense to repeat ourselves. Yet English is full of redundancies. We have "pleonasms" like *null and void, cease and desist,* and *for all intents and purposes,* and pointless redundancies like *advance planning.* And then there's the third-person singular suffix *-s,* which we use only when we can already tell from the subject that we have a third-person singular. The -*s* in *he buys,* relative to *they buy,* gives you no more information than if we just dropped the -*s* altogether and relied on the subject alone. The sentence *These three dogs are retrievers* conveys the notion of plurality not once but five times — in pluralizing the demonstrative pronoun (*these* as opposed to *this*), in the numeral (*three*), in the plural noun (*dogs* versus *dog*), in the verb (*are* versus *is*), and a final time in the final noun (*retrievers* versus *retriever*). In languages like Italian or Latin, which routinely omit subjects, a third-person plural marker makes sense; in English, which requires subjects, the third-person plural marker often adds nothing. Meanwhile, the phrase *John's picture,* which uses the possessive -*'s,* is ambiguous in at least three ways. Does it refer to a picture John took of someone else (say, his sister)? A photo that someone else (say, his sister) took of him? Or a picture of something else altogether (say, a blue-footed booby, *Sula nebouxii*), taken by someone else (perhaps a photographer from *National Geographic*), which John merely happens to own?

And then there's vagueness. In the sentence *It's warm outside,* there's no clear boundary between what counts as warm and what counts as not warm. Is it 70 degrees? 69? 68? 67? I can keep dropping degrees, but where do we draw the line? Or consider a word like *heap.* How many stones does it take to form a heap? Philosophers like to amuse themselves with the following mind-twister, known as a *sorites* (rhymes with *pieties*) paradox:

> Clearly, one stone does not make a heap. If one stone is not enough to qualify as a heap of stones, nor should two, since adding one stone to a pile that is not a heap should not turn that pile into a heap. And if two stones don't make a heap, three stones shouldn't either — by a logic that seemingly ought to extend to infinity. Working in the opposite direction, a man with 10,000 hairs surely isn't bald. But just as surely, plucking one hair from a man who is not bald should not produce a transition from not-

bald to bald. So if a man with 9,999 hairs cannot be judged to be bald, the same should apply to a man with 9,998. Following the logic to its extreme, hair by hair, we are ultimately unable even to call a man with zero hairs "bald."

If the boundary conditions of words were more precise, such reasoning (presumably fallacious) might not be so tempting.

Adding to the complication is the undeniable fact that languages just can't help but change over time. Sanskrit begat Hindi and Urdu; Latin begat French, Italian, Spanish, and Catalan. West Germanic begat Dutch, German, Yiddish, and Frisian. English, mixing its Anglo-Saxon monosyllables (*Halt!*) with its Greco-Latin impress-your-friends polysyllables *(Abrogate all locomotion!)*, is the stepchild of French and West Germanic, a little bit country, a little bit rock-and-roll.

Even where institutions like l'Académie française try to legislate language, it remains unruly. L'Académie has tried to bar from French such English-derived as *le hamburger, le drugstore, le week-end, le strip-tease, le pull-over, le tee-shirt, le chewing gum,* and *la cover-girl* — with no success whatsoever. With the rapid development of popular new techonology — such as iPods, podcasts, cell phones, and DVDs — the world needs new words every day.*

Most of us rarely notice the instability or vagueness of language, even when our words and sentences aren't precise, because we can decipher language by supplementing what grammar tells us with our *knowledge of the world*. But the fact that we *can* rely on something other than language — such as shared background knowledge — is no defense. When I "know what you mean" even though you haven't said it, language itself has fallen short. And when languages in general show evidence of these same problems, they reflect not only

---

*Perhaps even more galling to Franco-purists is that their own *fabrique de Nîmes* became known in English as "denim" — only to return to France as simply *les blue jeans* in the mother tongue. There are barbarians at the gate, and those barbarians are us.

cultural history but also the inner workings of the creatures who learn and use them.

Some of these facts about human language have been recognized for at least two millennia. Plato, for example, worried in his dialogue *Cratylus* that "the fine fashionable language of modern times has twisted and disguised and entirely altered the original meaning" of words. Wishing for a little more systematicity, he also suggested that "words should as far as possible resemble things . . . if we could always, or almost always, use likenesses, which are perfectly appropriate, this would be the most perfect state of language."

From the time of twelfth-century mystic Hildegard of Bingen, if not earlier, some particularly brave people have tried to do something about the problem and attempted to build more sensible languages from scratch. One of the most valiant efforts was made by English mathematician John Wilkins (1614–1672), who addressed Plato's concern about the systematicity of words. Why, for example, should cats, tigers, lions, leopards, jaguars, and panthers each be named differently, despite their obvious resemblance? In his 1668 work *An Essay Towards a Real Character and a Philosophical Language,* Wilkins sought to create a systematic "non-arbitrary" lexicon, reasoning that words ought to reflect the relations among things. In the process, he made a table of 40 major concepts, ranging from quantities, such as magnitude, space, and measure, to qualities, such as habit and sickness, and then he divided and subdivided each concept to a fine degree. The word *de* referred to the elements (earth, air, fire, and water), the word *deb* referred to fire, the first (in Wilkins's scheme) of the elements, *debα* to a part of fire, namely a flame, *deba* to a spark, and so forth, such that every word was carefully (and predictably) structured.

Most languages don't bother with this sort of order, incorporating new words catch-as-catch-can. As a consequence, when we English speakers see a rare word, say, *ocelot,* we have nowhere to start in determining its meaning. Is it a cat? A bird? A small ocean? Unless we speak Nahuatl (a family of native North Mexican languages that includes Aztec), from which the word is derived, we have no clue. Where Wilkins promised systematicity, we have only etymology, the history of a word's origin. An ocelot, as it happens, is a wild feline that gets its name from North Mexico; going further south, pumas are felines from Peru. The word *jaguar* comes from the Tupi language of Brazil. Meanwhile, the words *leopard, tiger,* and *panther* appear in ancient Greek. From the perspective of a child, each word is a fresh learning challenge. Even for adults, words that come up rarely are difficult to remember.

Among all the attempts at a perfect language, only one has really achieved any traction — Esperanto, created by one Ludovic Lazarus Zamenhof, born on December 15, 1859. Like Noam Chomsky, the father of modern linguistics, Zamenhof was son of a Hebrew scholar. By the time he was a teenager, little Ludovic had picked up French, German, Polish, Russian, Hebrew, Yiddish, Latin, and Greek. Driven by his love for language and a belief that a universal language could alleviate many a social ill, Zamenhof aimed to create one that could quickly and easily be acquired by any human being.

Saluton! Cu vi parolas Esperanton? Mia nomo estas Gary.
[Hello. Do you speak Esperanto? My name is Gary.]

Despite Zamenhof's best efforts, Esperanto is used today by only a few million speakers (with varying degrees of expertise), one tenth of 1 percent of the world's population. What makes one language more prevalent than another is mostly a matter of politics, money, and influence. French, once the most commonly spoken language in the West, wasn't displaced by English because English is better, but because Britain and the United States became more powerful and more influential than France. As the Yiddish scholar Max Weinrich put it, *"A shprakh iz a diyalekt mit an armey un a flot"* — "The only difference between a language and a dialect is an army and a navy."

With no nation-state invested in the success of Esperanto, it's perhaps not surprising that it has yet to displace English (or French, Spanish, German, Chinese, Japanese, Hindi, or Arabic, to name a few) as the most prevalent language in the world. But it is instructive nonetheless to compare it to human languages that emerged naturally. In some ways, Esperanto is a dream come true. For example, whereas German has a half-dozen different ways to form the plural, Esperanto has only one. Any language student would sigh with relief.

Still, Esperanto gets into some fresh troubles of its own. Because of its strict rules about stress (the penultimate syllable, always), there is no way to distinguish whether the word *senteme* is made up of sent + em + e ("feeling" + "a tendency toward" + adverbial ending) or sen + tem + e ("without" + "topic" + adverbial ending). Thus the sentence *La profesoro senteme parolis dum du horoj* could mean either "The professor spoke with feeling for two hours" or (horrors!) "The professor rambled on for two hours." The sentence *Estis batata la demono de la viro* is triply ambiguous; it can mean "The demon was beaten by the man," "The demon was beaten out of the man," or "The man's demon was beaten." Obviously, banishing irregularity is one thing, banishing ambiguity another.

Computer languages don't suffer from these problems; in Pascal, C, Fortran, or LISP, one finds neither rampant irregularity nor pervasive ambiguity — proof in principle that languages don't *have* to be ambiguous. In a well-constructed program, no computer ever wavers about what it should do next. By the very design of the languages in which they are written, computer programs are never at a loss.

Yet no matter how clear computer languages may be, nobody speaks C, Pascal, or LISP. Java may be the computer world's current lingua franca, but I surely wouldn't use it to talk about the weather. Software engineers depend on special word processors that indent, colorize, and keep track of their words and parentheses, precisely because the structure of computer languages seems so unnatural to the human mind.

To my knowledge, only one person ever seriously tried to construct an ambiguity-free, mathematically perfect human language, mathematically perfect not just in vocabulary but also in sentence construction. In the late 1950s a linguist by the name of James Cooke Brown constructed a language known as Loglan, short for "logical language." In addition to a Wilkins-esque systematic vocabulary, it includes 112 "little words" that govern logic and structure. Many of these little words have English equivalents (*tui,* "in general"; *tue,* "moreover"; *tai,* "above all"), but the really crucial words correspond to things like parentheses (which most spoken languages lack) and technical tools for picking out specific individuals mentioned earlier in the discourse. The English word *he,* for example, would be translated as *da* if it refers to the first singular antecedent in a discourse, *de* if it refers to the second, *di* if it refers to the third, *do* if it refers the fourth, and *du* if it refers to the fifth. Unnatural as this might seem, this system would banish considerable confusion about the antecedents of pronouns. (American Sign Language uses physical space to represent something similar, making signs in different places, depending on which entity is being referred to.) To see why this is useful, consider the English sentence *He runs and he walks.* It might describe a single person who runs and walks, or two different people, one running, the other walking; by contrast, in Loglan, the former would be rendered unambiguously as *Da prano i da dzoru,* the latter unambiguously as *Da prano i de dzoru.*

But Loglan has made even fewer inroads than Esperanto. Despite its "scientific" origins, it has no native speakers. On the Loglan website, Brown reports that at "The Loglan Institute . . . live-in apprentices learned the language directly from me (and I from them!), I am happy to report that sustained daily Loglan-only conversations lasting three-quarters of an hour or more were achieved," but so far as I know, nobody has gotten much further. For all its ambiguity and idiosyncrasy, English goes down much smoother for the human mind. We couldn't learn a perfect language if we tried.

As we have seen already, idiosyncrasy often arises in evolution when function and history clash, when good design is at odds with the raw materials already close at hand. The human spine, the panda's thumb (formed from a wrist bone) — these are ramshackle solutions that owe more to evolutionary inertia than to any principle of good design. So it is with language too.

In the hodgepodge that is language, at least three major sources of idiosyncrasy arise from three separate clashes: (1) the contrast between the way our ancestors made sounds and the way we would ideally like to make them, (2) the way in which our words build on a primate understanding of the world, and (3) a flawed system of memory that works in a pinch but makes little sense for language. Any one of these alone would have been enough to leave language short of perfection. Together, they make language the collective kluge that it is: wonderful, loose, and flexible, yet manifestly rough around the edges.

Consider first the very sounds of language. It's probably no accident that language evolved primarily as a medium of sound, rather than, say, vision or smell. Sound travels over reasonably long distances, and it allows one to communicate in the dark, even with others one can't see. Although much the same might be said for smell, we can modulate sound much more rapidly and precisely, faster than even the most sophisticated skunk can modulate odor. Speech is also faster than communicating by way of physical motion; it can flow at about twice the speed of sign language.

Still, if I were building a system for vocal communication from scratch, I'd start with an iPod: a digital system that could play back any sound equally well. Nature, in contrast, started with a breathing tube. Turning that breathing tube into a means of vocal production was no small feat. Breathing produces air, but sound is modulated air, vibrations produced at just the right sets of frequencies. The Rube

Goldberg–like vocal system consists of three fundamental parts: respiration, phonation, and articulation.

Respiration is just what it sounds like. You breathe in, your chest expands; your chest compresses, and a stream of air comes out. That stream of air is then rapidly divided by the vocal folds into smaller puffs of air (phonation), about 80 times a second for a baritone like James Earl Jones, as much as 500 times per second for a small child. From there, this more-or-less constant sound source is filtered, so that only a subset of its many frequencies makes it through. For those who like visual analogies, imagine producing a perfect white light and then applying a filter, so that only part of the spectrum shines through. The vocal tract works on a similar "source and filter" principle. The lips, the tip of the tongue, the tongue body, the velum (also known as the soft palate), and the glottis (the opening between the vocal folds) are known collectively as articulators. By varying their motions, these articulators shape the raw sound stream into what we know as speech: you vibrate your vocal cords when you say "bah" but not "pah"; you close your lips when say "mah" but move your tongue to your teeth when you say "nah."

Respiration, phonation, and articulation are not unique to humans. Since fish walked the land, virtually all vertebrates, from frogs to birds to mammals, have used vocally produced sound to communicate. Human evolution, however, depended on two key enhancements: the lowering of our larynx (not unique to humans but very rare elsewhere in the animal kingdom) and increased control of the ensemble of articulators that shape the sound of speech. Both have consequences.

Consider first the larynx. In most species, the larynx consists of a single long tube. At some point in evolution, our larynx dropped down. Moreover, as we changed posture and stood upright, it took a 90-degree turn, dividing into two tubes of more or less equal length, which endowed us with considerably more control of our vocalizations — and radically increased our risk of choking. As first noted by

Darwin, "Every particle of food and drink which we swallow has to pass over the orifice of the trachea, with some risk of falling into the lungs" — something we're all vulnerable to.*

Maybe you think the mildly increased risk of choking is a small price to pay, maybe you don't. It certainly didn't *have* to be that way; breathing and talking could have relied on different systems. Instead, our propensity for choking is one more clear sign that evolution tinkered with what was already in place. The result is a breathing tube that does double duty as a vocal tract — in occasionally fatal fashion.

In any event, the descended larynx was only half the battle. The real entrée into speech came from significantly increased control over our articulators. But here too the system is a bit of a kluge. For one thing, the vocal tract lacks the elegance of the iPod, which can play back more or less any sound equally well, from Moby's guitars and flutes to hip-hop's car crashes and gunshots. The vocal tract, in contrast, is tuned only to words. All the world's languages are drawn from an inventory of 90 sounds, and any particular language employs no more than half that number — an absurdly tiny subset when you think of the many distinct sounds the ear can recognize.

Imagine, for example, a human language that would refer to something by reproducing the sound it makes. I'd refer to my favorite canine, Ari, by reproducing his woof, not by calling him a dog. But the three-part contraption of respiration, phonation, and articula-

---

*According to a recent article in *The New Yorker,* were it not for the Heimlich maneuver, all of the following people might have choked: Cher (vitamin pill), Carrie Fisher (Brussels sprout), Dick Vitale (melon), Ellen Barkin (shrimp), and Homer Simpson (doughnut). The article continued with a list of "heroes," celebrities who saved others from certain death: "Tom Brokaw (John Chancellor, Gouda cheese), Verne Lundqvist (Pat Haden, broccoli), Pierce Brosnan (Halle Berry, fruit), Justin Timberlake (a friend, nuts), Billy Bob Thornton (his potbellied pig Albert, chicken Marsala)." Especially eerie was the tale of actor Mandy Patinkin, saved from a caesar salad just three weeks after wrapping a film called — I kid you not — *The Choking Man.*

tion can only do so much; even where languages allegedly refer to objects by their sounds — the phenomenon known as onomatopoeia — the "sounds" we refer to sound like, well, *words*. *Woof* is a perfectly well formed English word, a cross between, say, *wool* and *hoof*, but not a faithful reproduction of Ari's vocalization (nor that of any other dog). And the comparable words in other languages each sound different, none exactly like a woof or a bark. French dogs go *ouah, ouah*, Albanian dogs go *ham, ham*, Greek dogs go *gav, gav*, Korean dogs go *mung, mung*, Italian dogs go *bau, bau*, German dogs *wau, wau*: each language creates the sound in its own way. Why? Because our vocal tract is a clumsy contraption that is good for making the sounds of speech — and little else.

Tongue-twisters emerge as a consequence of the complicated dance that the articulators perform. It's not enough to close our mouth or move our tongue in a basic set of movements; we have to coordinate each one in precisely timed ways. Two words can be made up of exactly the same physical motions performed in a slightly different sequence. *Mad* and *ban*, for example, each require the same four crucial movements — the velum (soft palate) widens, the tongue tip moves toward alveolar closure, the tongue body widens in the pharynx, and the lips close — but one of those gestures is produced early in one word (*mad*) and late in another (*ban*). Problems occur as speech speeds up — it gets harder and harder to get the timing right. Instead of building a separate timer (a clock) for each gesture, nature forces one timer into double (or triple, or quadruple) duty.

And that timer, which evolved long before language, is really good at only very simple rhythms: keeping things either exactly in phase (clapping) or exactly out of phase (alternating steps in walking, alternating strokes in swimming, and so forth). All that is fine for walking or running, but not if you need to perform an action with a more complex rhythm. Try, for example, to tap your right hand at twice the rate of your left. If you start out slow, this should be easy. But now gradually increase the tempo. Sooner or later you will find that the rhythm of your tapping will break down (the technical term is *devolve*) from a ratio of 2:1 to a ratio of 1:1.

Which returns us to tongue-twisters. Saying the words *she sells* properly involves a challenging coordination of movements, very much akin to tapping at the 2:1 ratio. If you first say the words *she* and *sells* aloud, slowly and separately, you'll realize that the /s/ and /sh/ sounds have something in common — a tongue-tip movement — but only /sh/ also includes a tongue-body gesture. Saying *she sells* properly thus requires coordinating two tongue-tip gestures with one tongue-body gesture. When you say the words slowly, everything is okay, but say them fast, and you'll stress the internal clock. The ratio eventually devolves to 1:1, and you wind up sticking in a tongue-body gesture for every tongue-tip gesture, rather than every other one. Voilà, *she sells* has become *she shells*. What "twists" your tongue, in short, is not a muscle but a limitation in an ancestral timing mechanism.

The peculiar nature of our articulatory system and how it evolved, leads to one more consequence: the relation between sound waves and phonemes (the smallest distinct speech sounds, such as /s/ and /ā/) is far more complicated than it needs to be. Just as our pronunciation of a given sequence of letters depends on its linguistic context (think of how you say *ough* when reading the title of Dr. Seuss's book *The Tough Coughs As He Ploughs the Dough*), the way in which we produce a particular linguistic element depends on the sounds that come before it and after it. For example, the sound /s/ is pronounced in one way in the word *see* (with spread lips) but in another in the word *sue* (with rounded lips). This makes learning to talk a lot more work than it might otherwise be. (It's also part of what makes computerized voice-recognition a difficult problem.)

Why such a complex system? Here again, evolution is to blame; once it locked us into producing sounds by articulatory choreography, the only way to keep up the speed of communication was to cut corners. Rather than produce every phoneme as a separate, distinct element (as a simple computer modem would), our speech system starts preparing sound number two while it's still working on sound number one. Thus, before I start uttering the *h* in *happy,* my tongue is already scrambling into position in anticipation of the *a.* When I'm working on *a,* my lips are already getting ready for the *pp,* and when I'm on *pp,* I'm moving my tongue in preparation for the *y.*

This dance keeps the speed up, but it requires a lot of practice and can complicate the interpretation of the message.* What's good for muscle control isn't necessarily good for a listener. If you should mishear John Fogerty's "There's a bad moon on the rise" as "There's a bathroom on the right,"† so be it. From the perspective of evolution, the speech system, which works most of the time, is good enough, and that's all that matters.

Curmudgeons of every generation think that their children and grandchildren don't speak properly. Ogden Nash put it this way in 1962, in "Laments for a Dying Language":

> Coin brassy words at will, debase the coinage;
> We're in an if-you-cannot-lick-them-join age,
> A slovenliness provides its own excuse age,
> Where usage overnight condones misusage.
> Farewell, farewell to my beloved language,
> Once English, now a vile orangutanguage.

Words in computer languages are fixed in meaning, but words in human languages change constantly; one generation's *bad* means

---

*Co-articulation did not evolve exclusively for use in speech; we see the same principle at work in skilled pianists (who prepare for thumb-played notes about two notes before they play them), skilled typists, and major league baseball pitchers (who prepare the release of the ball well before it occurs).

†Or Jimi Hendrix's "Excuse me while I kiss the sky" for "Excuse me while I kiss this guy." If you, like me, get a kick out of these examples, Google for the term *Mondegreen* and find oodles more.

"bad," and the next generation's *bad* means "good." Why is it that languages can change so quickly over time?

Part of the answer stems from how our *prelinguistic* ancestors evolved to think about the world: not as philosophers or mathematicians, brimming with precision, but as animals perpetually in a hurry, frequently settling for solutions that are "good enough" rather than definitive.

Take, for example, what might happen if you were walking through the Redwood Forest and saw a tree trunk; odds are, you would conclude that you were looking at a tree, even if that trunk happened to be so tall that you couldn't make out any leaves above. This habit of making snap judgments based on incomplete evidence (no leaves, no roots, just a trunk, and still we conclude we've seen a tree) is something we might call a logic of "partial matching."

The logical antithesis, of course, would be to wait until we'd seen the whole thing; call that a logic of "full matching." As you can imagine, he who waits until he's seen the whole tree would never be wrong, but also risks missing a lot of bona fide foliage. Evolution rewarded those who were swift to decide, not those who were too persnickety to act.

For better or worse, language inherited this system wholesale. You might think of a chair, for instance, as something with four legs, a back, and a horizontal surface for sitting. But as the philosopher Ludwig Wittgenstein (1889–1951) realized, few concepts are really defined with such precision. Beanbag chairs, for example, are still considered chairs, even though they have neither an articulated back nor any sort of legs.

I call my cup of water a glass even though it's made of plastic; I call my boss the chair of my department even though so far as I can tell she merely sits in one. A linguist or phylogenist uses the word *tree* to refer to a diagram on a page simply because it has branching structures, not because it grows, reproduces, or photosynthesizes. A head is the topside of a penny, the tail the bottom, even though the top has no more than a picture of a head, the bottom not a fiber of a wagging tail. Even the slightest fiber of connection suffices, precisely because words are governed by an inherited, ancestral logic of partial matches.*

Another idiosyncrasy of language, considerably more subtle, has to do with words like *some, every,* and *most,* known to linguists as "quantifiers" because they quantify, answering questions like "How much?" and "How many?": *some water, every boy, most ideas, several movies.*

The peculiar thing is that in addition to quantifiers, we have another whole system that does something similar. This second system traffics in what linguists call "generics," somewhat vague, generally accurate statements, such as *Dogs have four legs* or *Paperbacks are cheaper than hardcovers.* A perfect language might stick only to the first system, using explicit quantifiers rather than generics. An explicitly quantified sentence such as *Every dog has four legs* makes a nice, strong, clear statement, promising no exceptions. We know how to figure out whether it is true. Either all the dogs in the world have four legs, in which case the sentence is true, or at least one dog lacks four legs, in which case the sentence is false — end of story. Even a quantifier like *some* is fairly clear in its application; *some* has to mean more than one, and (pragmatically) ought not to mean *every.*

Generics are a whole different ball game, in many ways much less precise than quantifiers. It's just not clear how many dogs have to have four legs before the statement *Dogs have four legs* can be considered true, and how many dogs would have to exhibit three legs before we'd decide that the statement is false. As for *Paperbacks are cheaper than hardcovers,* most of us would accept the sentence as true as a general rule of thumb, even if we knew that lots of individual paperbacks (say, imports) are more expensive than many individual hardcovers (such as discounted bestsellers printed in large quantities). We agree with the statement *Mosquitoes carry the West Nile vi-*

---

*Is that good or bad? That depends on your point of view. The logic of partial matches is what makes languages sloppy, and, for better or worse, keeps poets, stand-up comedians, and linguistic curmudgeons gainfully employed. ("Didja ever notice that a near-miss isn't a miss at all?")

rus, even if only (say) 1 percent of mosquitoes carry the virus, yet we wouldn't accept the statement *Dogs have spots* even if all the dalmatians in the world did.

Computer-programming languages admit no such imprecision; they have ways of representing formal quantifiers ([DO THIS THING REPEATEDLY UNTIL EVERY DATABASE RECORD HAS BEEN EXAMINED]) but no way of expressing generics at all. Human languages are idiosyncratic — and verging on redundant — inasmuch as they routinely exploit both systems, generics and the more formal quantifiers.

Why do we have both systems? Sarah-Jane Leslie, a young Princeton philosopher, has suggested one possible answer. The split between generics and quantifiers may reflect the divide in our reasoning capacity, between a sort of fast, automatic system on the one hand and a more formal, deliberative system on the other. Formal quantifiers rely on our deliberative system (which, when we are being careful, allows us to reason logically), while generics draw on our ancestral reflexive system. Generics are, she argues, essentially a linguistic realization of our older, less formal cognitive systems. Intriguingly, our sense of generics is "loose" in a second way: we are prepared to accept as true generics like *Sharks attack bathers* or *Pit bulls maul children* even though the circumstances they describe are statistically very rare, provided that they are vivid or salient — just the kind of response we might expect from our automatic, less deliberative system.

Leslie further suggests that generics seem to be learned first in childhood, before formal quantifiers; moreover, they may have emerged earlier in the development of language. At least one contemporary language (Piraha, spoken in the Amazon Basin) appears to employ generics but not formal quantifiers. All of this suggests one more way in which the particular details of human languages depend on the idiosyncrasies of how our mind evolved.

For all that, I doubt many linguists would be convinced that language is truly a kluge. Words are one thing, sentences another; even if words are clumsy, what linguists really want to know about is *syntax,* the glue that binds words together. Could it be that words are a mess, but grammar is different, a "near-perfect" or "optimal" system for connecting sound and meaning?

In the past several years, Noam Chomsky, the founder and leader of modern linguistics, has taken to arguing just that. In particular, Chomsky has wondered aloud whether language (by which he means mainly the syntax of sentences) might come close "to what some super-engineer would construct, given the conditions that the language faculty must satisfy." As linguists like Tom Wasow and Shalom Lappin have pointed out, there is considerable ambiguity in Chomsky's suggestion. What would it mean for a language to be perfect or optimal? That one could express anything one might wish to say? That language is the most efficient possible means for obtaining what one wants? Or that language was the most logical system for communication anyone could possibly imagine? It's hard to see how language, as it now stands, can lay claim to such grand credentials. The ambiguity of language, for example, seems unnecessary (as computers have shown), and language works in ways neither logical nor efficient (just think of how much extra effort is often required in order to clarify what our words mean). If language were a perfect vehicle for communication, infinitely efficient and expressive, I don't think we would so often need "paralinguistic" information, like that provided by gestures, to get our meaning across.

As it turns out, Chomsky actually has something different in mind. He certainly doesn't think language is a perfect tool for communication; to the contrary, he has argued that it is a mistake to think of language as having evolved "for" the purposes of communication at all. Rather, when Chomsky says that language is nearly optimal, he seems to mean that its formal structure is surprisingly *elegant,* in the same sense that string theory is. Just as string theorists conjecture that the complexity of physics can be captured by a small set of basic laws, Chomsky has, since the early 1990s, been trying to capture what he sees as the superficial complexity of language with a small set of laws.* Building on that idea, Chomsky and his collaborators have gone so far as to suggest that language might be a kind of "optimal solution . . . [to] the problem of linking the sensory-motor and conceptual-intentional systems" (or, roughly, connecting sound and meaning). They suggest that language, despite its obvious complexity, might have required only a *single* evolutionary advance beyond our inheritance from ancestral primates, namely, the introduction of a device known as "recursion."

Recursion is a way of building larger structures out of smaller structures. Like mathematics, language is a potentially infinite system. Just as you can always make a number bigger by adding one (a trillion plus one, a googleplex plus one, and so forth), you can always make a sentence longer by adding a new clause. My favorite example comes from Maxwell Smart on the old Mel Brooks TV show *Get Smart:* "Would you believe that I know that you know that I know that you know where the bomb is hidden?" Each additional clause requires another round of recursion.

There's no doubt that recursion — *or something like it* — is central to human language. The fact that we can put together one small bit of structure (*the man*) with another (*who went up the hill*) to form a more complex bit of structure (*the man who went up the hill*) allows us to create arbitrarily complex sentences with terrific precision (*The man with the gun is the man who went up the hill, not the man who*

---

*Although I have long been a huge fan of Chomsky's contributions to linguistics, I have serious reservations about this particular line of work. I'm not sure that elegance really works in physics (see Lee Smolin's recent book *The Trouble with Physics*), and in any case, what works for physics may well not work for linguistics. Linguistics, after all, is a property of biology — the biology of the human brain — and as the late Francis Crick once put it, "In physics, they have laws; in biology, we have gadgets." So far as we know, the laws of physics have never changed, from the moment of the big bang onward, whereas the details of biology are constantly in flux, evolving as climates, predators, and resources, change. As we have seen so many times, evolution is often more about alighting on something that happens to work than what might in principle work best or most elegantly; it would be surprising if language, among evolution's most recent innovations, was any different.

drove the getaway car). Chomsky and his colleagues even have suggested that recursion might be "the only uniquely human component of the faculty of language."

A number of scholars have been highly critical of that radical idea. Steven Pinker and the linguist Ray Jackendoff have argued that recursion might actually be found in other aspects of the mind (such as the process by which we recognize complex objects as being composed of recognizable subparts). The primatologist David Premack, meanwhile, has suggested that although recursion is a hallmark of human language, it is scarcely the *only* thing separating human language from other forms of communication. As Premack has noted, it's not as if chimpanzees can speak an otherwise humanlike language that lacks recursion (which might consist of language minus complexities such as embedded clauses).* I'd like to go even further, though, and take what we've learned about the nature of evolution and humans to turn the whole argument on its head.

The sticking point is what linguists call syntactic trees, diagrams like this:

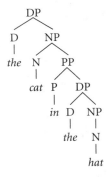

---

*In a hypothetical recursion-free language, you might, for example, be able to say "Give me the fruit" and "The fruit is on the tree," but not the more complex expression "Give me the fruit that is hanging on the tree that is missing a branch." The words "that is hanging on the tree that is missing a branch" represent an embedded clause itself containing an embedded clause.

Small elements can be combined to form larger elements, which in turn can be combined into still larger elements. There's no problem *in principle* with building such things — computers use trees, for example, in representing the directory, or "folder" structures, on a hard drive.

But, as we have seen time and again, what is natural for computers isn't always natural for the human brain: building a tree would require a precision in memory that humans just don't appear to have. Building a tree structure with postal-code memory is trivial, something that the world's computer programmers do many times a day. But building a tree structure out of *contextual memory* is a totally different story, a kluge that kind of works and kind of doesn't.

Working with simple sentences, we're usually fine, but our capacity to understand sentences can easily be compromised. Take, for example, this short sentence I mentioned in the opening chapter:

People people left left.

Here's a slightly easier variant:

Farmers monkeys fear slept.

Four words each, but enough to boggle most people's mind. Yet both sentences are perfectly grammatical. The first means that some set of people who were abandoned by a second group of people themselves departed; the second one means, roughly, "There is a set of farmers that the monkeys fear, and that set of farmers slept; the farmers that the monkeys were afraid of slept." These kinds of sentences — known in the trade as "center embeddings" (because they bury one clause directly in the middle of another) — are difficult, I submit, precisely because evolution never stumbled on proper tree structure.*

---

*Recursion can actually be divided into two forms, one that requires a stack and one that doesn't. The one that doesn't is easy. For example, we have no trouble with sentences like *This is the cat that bit the rat that chased the mouse*, which are complex but (for technical reasons) can be parsed without a stack.

Here's the thing: in order to interpret sentences like these and fully represent recursion (another classic is *The rat the cat the mouse chased bit died*), we would need to keep track of each noun and each verb, and at the same time hold in mind the connections between them and the clauses they form. Which is just what grammatical trees are *supposed* to do.

The trouble is, to do that would require an *exact* memory for the structures and words that have just been said (or read). And that's something our postal-code-free memories just aren't set up to do. If I were to read this book aloud and suddenly, without notice, stop and ask you to repeat the last sentence you heard — you probably couldn't. You'd likely remember the *gist* of what I had said, but the exact wording would almost surely elude you.*

As a result, efforts to keep track of the structure of sentences becomes a bit like efforts to reconstruct the chronology of a long-ago sequence of events: clumsy, unreliable, but better than nothing. Consider, for example, a sentence like *It was the banker that praised the barber that alienated his wife that climbed the mountain.* Now, quick: was the mountain climbed by the banker, the barber, or his wife? A computer-based parser would have no trouble answering this question; each noun and each verb would be slotted into its proper place in a tree. But many human listeners end up confused. Lacking any hint of memory organized by location, the best we can do is *approximate* trees, clumsily kluging them together out of contextual memory. If we receive enough distinctive clues, it's not a problem, but when the individual components of sentences are similar enough to confuse, the whole edifice comes tumbling down.†

Perhaps the biggest problem with grammar is not the trouble we have in constructing trees, but the trouble we have in producing sentences

---

*Perhaps the most extreme version of remembering only the gist was Woody Allen's five-word summary of *War and Peace:* "It was about some Russians."

†The problem with trees is much the same as the problem with keeping tracking of our goals. You may recall, from the chapter on memory, the example of what some-

that are certain to be parsed as we intend them to be. Since our sentences are clear to *us*, we assume they are clear to our listeners. But often they're not; as engineers discovered when they started trying to build machines to understand language, a significant fraction of what we say is quietly ambiguous.*

Take, for example, this seemingly benign sentence: *Put the block in the box on the table.* An ordinary sentence, but it can actually mean two things: a request to put a particular block that happens to be in a box onto the table, or a request to take some block and put it into a particular box that happens to be on the table. Add another clause, and we wind up with four possibilities:

- Put the block [(in the box on the table) in the kitchen].
- Put the block [in the box (on the table in the kitchen)].
- Put [the block (in the box) on the table] in the kitchen.
- Put (the block in the box) (on the table in the kitchen).

Most of the time, our brain shields us from the complexity, automatically doing its best to reason its way through the possibilities. If we hear *Put the block in the box on the table,* and there's just one block, we don't even notice the fact that the sentence could have meant something else. Language alone doesn't tell us that, but we are clever enough to connect what we hear with what it might mean. (Speakers also use a range of "paralinguistic" techniques, like point-

---

times happens when we plan to stop at the grocery store after work (and instead "autopilot" our way home, sans groceries). In a computer, both types of problems — tracking goals and tracking trees — are typically solved by using a "stack," in which recent elements temporarily take priority over stored ones; but when it comes to humans, our lack of postal-code memory leads to problems in both cases.

As it happens, there are actually two separate types of recursion, one that requires stacks and one that doesn't. It is precisely the ones that do require stacks that tie us in knots.

*According to legend, the first machine translation program was given the sentence "The flesh is weak, but the spirit is willing." The translation (into Russian) was then translated back into English, yielding, "The meat is spoiled, but the vodka is good."

ing and gesturing, to supplement language; they can also look to their listeners to see if they appear to understand.)

But such tricks can take us only so far. When we are stuck with inadequate clues, communication becomes harder, one reason that emails and phone calls are more prone to misunderstandings than face-to-face communication is. And even when we speak directly to an audience, if we use ambiguous sentences, people may just not notice; they may think they've understood even when they haven't really. One eye-opening study recently asked college students to read aloud a series of grammatically ambiguous* sentences like *Angela shot the man with the gun* (in which the gun might have been either Angela's murder weapon or a firearm the victim happened to be carrying). They were warned in advance that the sentences were ambiguous and permitted to use as much stress (emphasis) on individual words as they liked; the question was whether they could *tell* when they successfully put their meaning across. It turns out that most speakers were lousy at the task and had no clue about how bad they were. In almost half the cases in which subjects thought that they had successfully conveyed a given sentence's meaning, they were actually misunderstood by their listeners.† (The listeners weren't much better, frequently assuming they'd understood when they hadn't.)

---

*Ambiguity comes in two forms, lexical and syntactic. Lexical ambiguity is about the meanings of individual words; I tell you to go have a ball, and you don't know whether I mean a good time, an elaborate party, or an object for playing tennis. Syntactic (or grammatical) ambiguity, in contrast, is about sentences like *Put the block on the box in the table,* that have structures which could be interpreted in more than one way. Classic sentences like *Time flies like an arrow* are ambiguous in both ways; without further context, *flies* could be a noun or a verb, *like* a verb or a comparative, and so forth.

†In a perfect language, in an organism with properly implemented trees, this sort of inadvertent ambiguity wouldn't be a problem; instead we'd have the option of using what mathematicians use: parentheses, which are basically symbols that tell us how to group things. $(2 \times 3) + 2 = 8$, while $2 \times (3 + 2) = 10$. We'd be able to easily articulate the difference between *(Angela shot the man) with the gun* and *Angela shot (the man with the gun).* As handy as parentheses would be, they, too, are missing in action because the species-wide lack of postal code memory.

Indeed, a certain part of the work that professional writers must do (especially if they write nonfiction) is compensate for language's limitations, to scan carefully to make sure that there's no vague *he* that could refer to either the farmer or his son, no misplaced commas, no dangling (or squinting) modifiers, and so forth. In Robert Louis Stevenson's words, "The difficulty of literature is not to write, but to write what you mean." Of course, sometimes ambiguity is deliberate, but that's a separate story; it's one thing to leave a reader with a vivid sense of a difficult decision, another to accidentally leave a reader confused.

Put together all these factors — inadvertent ambiguity, idiosyncratic memory, snap judgments, arbitrary associations, and a choreography that strains our internal clocks — and what emerges? Vagueness, idiosyncrasy, and a language that is frequently vulnerable to misinterpretation — not to mention a vocal apparatus more byzantine than a bagpipe made up entirely of pipe cleaners and cardboard dowels. In the words of the linguist Geoff Pullum, "The English language is, in so many ways, a flawed masterpiece of evolution, loaded with rough bits, silly design oversights, ragged edges, stupid gaps, and malign and perverted irregularities."

As the psycholinguist Fernanda Ferreira has put it, language is "good enough," not perfect. Most of the time we get things right, but sometimes we are easily confused. Or even misled. Few people, for example, scarcely notice that something's amiss when you ask them, "How many animals did Moses bring onto the ark?"* Even fewer realize that a sentence like *More people have been to Russia than I have* is either (depending on your point of view) ungrammatical or incoherent.

If language were designed by an intelligent engineer, interpreters would be out of a job, and Berlitz's language schools would be drive-

---

*People hear the words *animals* and *ark,* and fail to notice that the question is about Moses rather than Noah.

thrus, no lifetime commitment required. Words would be systematically related to one another, and phonemes consistently pronounced. You could tell all those voice-activated telephone menu systems exactly where you wanted them to go — and be assured they'd understand the message. There would be no ambiguity, no senseless irregularity. People would say what they mean and mean what they say. But instead, we have slippage. Our thoughts get stuck on the tip of the tongue when we can't recall a specific word. Grammar ties us in knots (is it *The keys to the cabinet are* ... or *The keys to the cabinet is* ... Oh never mind ...). Syntax on the fly is hard.

This is not to say that language is terrible, only that it could, with forethought, have been even better.

The rampant confusion that characterizes language is not, however, without its logic: the logic of evolution. We co-articulate, producing speech sounds differently, depending on the context, because we produce sound not by running a string of bits through a digital amplifier to electromagnetically driven speakers but by thrashing our tongues around three-dimensional cavities that originated as channels for digestion, not communication. Then, as *She sells seashells by the seashore*, our tortured tongues totally trip. Why? Because language was built, rapidly, on a haphazard patchwork of mechanisms that originally evolved for other purposes.

6

PLEASURE

Happiness is a warm puppy.

— CHARLIE BROWN

Happiness is a warm gun.

— THE BEATLES

To each his own.

— Traditional saying

WOE BETIDE THE HUMAN BEING who doesn't know what happiness is; yet woe to the writer who tries to define it. Warm guns and warm puppies are merely *examples* of happiness, not definitions of it.

My dictionary defines happiness as "pleasure" — and pleasure as a feeling of "happy satisfaction and enjoyment." As if that weren't circular enough, when I turn to the word *feeling*, I find that a feeling is defined as "a perceived emotion" while an emotion is defined as a "strong feeling."

No matter. As the Supreme Court justice Potter Stewart famously said about pornography (as opposed to art), it is hard to define, but "I know it when I see it." Happiness may mean sex, drugs, and rock-and-roll, the roar of the crowd, the satisfaction of a job well done, good food, good drink, and good conversation — not to mention what psychologist Mihaly Csikszentmihalyi calls a state of "flow," of being so absorbed in something you do well that you scarcely notice the passage of time. At the risk of offending hard-nosed philosophers everywhere, I propose to leave it at that. For my money, the real question is not how we define happiness, but *why*, from the perspective of evolution, humans care about it at all.

At first glance, the answer seems obvious. The standard story is that happiness evolved in part to guide our behavior. In the words of the noted evolutionary psychologist Randolph Nesse, "Our brains could have been wired so that [eating] good food, [having] sex, being the object of admiration, and observing the success of one's children were all aversive experiences [but] any ancestor whose brain was so wired would probably not have contributed much to the gene pool that makes human nature what it is now." Pleasure is our guide, as Freud (and long before him, Aristotle) noted, and without it, the species wouldn't propagate.*

That much seems true. In keeping with the notion that pleasure serves as our guide, we automatically (and often unconsciously) sort just about everything we see into the categories "pleasant" and "unpleasant." If I show you a word like *sunshine* and then ask you to decide, as quickly as possible, whether the word *wonderful* is positive, you'll respond faster than you would if shown an unpleasant word instead (say, *poison* instead of *sunshine*). Cognitive psychologists call this accelerated response a positive priming effect; it means that we constantly and automatically categorize everything we encounter as good or bad.

This sort of automatic evaluation, largely the domain of the reflexive system, is remarkably sophisticated. Take, for example, the word *water;* is it pleasant? Depends on how thirsty you are. And sure enough, studies show that thirsty people show a bigger positive prim-

---

*That said, when it comes to reasons for having sex, pleasure and reproduction are, at least for humans, just two motivations among many. The most comprehensive survey ever conducted, reported recently in the *Archives of Sexual Behavior,* listed a grand total of 237. Seeking pleasure and making babies were definitely on the list, but so were "it seemed like good exercise," "I wanted to be popular," "I was bored," "I wanted to say thank you," and "I wanted to feel closer to God." One way or another, 96 percent of American adults have at least once found some reason for having sex.

ing effect for the word *water* than people who are well hydrated. This occurs over the course of a few milliseconds, allowing pleasure to serve as a moment-by-moment guide. Similar findings — and this is the scary part — apply even in our attitudes toward other people: the more we need them, the more we like them. (To rather cynically paraphrase an old expression, from the perspective of the subconscious, a friend we need is a friend perceived.)

But the simple idea that "if it feels good, it must have been good for our ancestors" runs into trouble pretty quickly. To start with, many — arguably, most — things that give us pleasure don't actually do much for our genes. In the United States, the average adult spends nearly a third of his or her waking hours on leisure activities such as television, sports, drinking with friends — pursuits that may have little or no direct genetic benefit. Even sex, most of the time for most people, is recreational, not procreational. When I spend $100 on a meal at Sushi Samba, my favorite restaurant du jour, I don't do it because it will increase the number of kids I have, or because eating Peruvian-Japanese fusion food is the cheapest (or even most nutritious) way to fill my belly. I do it because, well, I *like* the taste of yellowtail ceviche — even if, from the perspective of evolution, my dining bills are a waste of precious financial resources.

A Martian looking down on planet Earth might note all this with puzzlement. Why do humans fool around so much when there is, inevitably, work to be done? Although other species have been known to play, no other species goofs around so much, or in so many ways. Only a few other species seem to spend much time having nonprocreative sex, and none (outside labs run by inquisitive humans) watch television, go to rock concerts, or play organized sports. Which raises the question, is pleasure really an ideal adaptation, or (with apologies to Shakespeare) is there something klugey in Denmark?

Aha, says our Martian to itself; humans are no longer slaves to their genes. Instead of engaging in the activities that would yield the most copies of their genes, humans are trying to maximize something else, something more abstract — call it "happiness" — which appears to be a measure of factors such as a human's general well-being, its level of success, its perceived control over its own life, and how well it is regarded by its peers.

At which point, our Martian friend would be even more confused. If people are trying to maximize their overall well-being, why do they do so many things that in the long run yield little or no lasting happiness?

Perhaps nothing would puzzle this Martian more than the enormous amount of time that many people spend watching television. In America, the average is 2–4 hours *per day*. If you consider that the average person is awake for only 16, and at work for at least 8, that's a huge proportion of the average person's discretionary time. Yet day after day, audiences watch show after show, most containing either stories of dubious quality about fictional people or heavily edited "reality" portraits of people in improbable situations that the average viewer is unlikely ever to meet. (Yes, public television airs some great documentaries, but they never draw the ratings of *Law & Order, Lost,* or *Survivor.*) And here's the kicker: hard-core television watchers are, on average, less happy than those who watch only a little television. All that TV viewing might convey some sort of short-term benefit, but in the long term, an hour spent watching television is an hour that could have been spent doing something else — exercising, working on hobbies, caring for children, helping strangers, or developing friendships.

And then there are, of course, chemical substances deliberately designed to shortcut the entire machinery of reward, directly stimulating the pleasure parts of the brain (for example, the nucleus accumbens). I'm speaking, of course, of alcohol, nicotine, and drugs like cocaine, heroin, and amphetamines. What is remarkable about these substances is not the fact that they exist — it would be almost impossible to build a chemically based brain that *wouldn't* be vulnerable to the machinations of clever chemists — but rather the extent to which people use them, even when they realize that the long-term consequences may be life threatening. The writer John Cheever, for example, once wrote, "Year after year I read in [my journals] that I am drinking too much . . . I waste more days, I suffer deep pangs of guilt, I wake up at three in the morning with the feelings of a temperance worker. Drink, its implements, environments, and effects all seem disgusting. And yet each noon I reach for the whiskey bottle."

As one psychologist put it, addictions can lead people down a "primrose path" in which decisions made in the moment seem — from the strict perspective of temporary happiness — to be rational, even though the long-term consequences are often devastating.

Even sex has its puzzling side. That sex is enjoyable is perhaps no surprise: if sex weren't fun for our ancestors, we simply wouldn't be here. Sex is, after all, the royal road to conception, and without conception there would be no life. Without life, there would be no reproduction, and legions of "selfish genes" would be out of work. It seems like a no-brainer that creatures that enjoy sex (or at least are driven toward it) will outpropagate those that do not.

But having a taste for sex is not the same as pursuing it nonstop, to the virtual exclusion of anything else. We all know stories of politicians, priests, and plain ordinary folk who destroyed their life in relentless pursuit of sex. Might a Martian question whether our contemporary need for sex is as miscalibrated as our need for sugar, salt, and fat?

The Martian would eventually come to realize that although the core notion of pleasure as motivator makes a good deal of sense, the pleasure system as a whole is a kluge, from top to bottom. If pleasure is supposed to guide us to meet the needs of our genes, why do we humans fritter away so much of our time in activities that don't advance those needs? Sure, some men may skydive to impress the ladies, but many of us ski, snowboard, or drive recklessly even when nobody else is watching. When such a large part of human activity does something that *risks* "reproductive fitness," there must be some explanation.

And indeed there is, but it's not about minds that are optimal, but about minds that are clumsy. The first reason should, by now, seem familiar: the neural hardware that governs pleasure is, like much of the rest of the human mind, split in two: some of our pleasure (like, perhaps, the sense of accomplishment we get from a job well done) derives from the deliberative system, but most of it doesn't. Most pleasure springs from the ancestral reflexive system, which, as we have seen, is rather shortsighted, and the weighting between the two systems still favors the ancestral. Yes, I may get a slight sense of satisfaction if I waive my opportunity to eat that crème brûlée, but that satisfaction would almost certainly pale in comparison to the kick, however brief, that I would get from eating it.* My genes would be better off if I skipped dessert — my arteries might stay open that much longer, allowing me to gather more income and take better care of my future offspring — but those very genes, due to their lack of foresight, left me with a brain that lacks the wisdom to consistently outwit the animalistic parts of my brain, which are a holdover from an earlier era.

The second reason is more subtle: our pleasure center wasn't built for creatures as expert in culture and technology as we are; most of the mechanisms that give us pleasure are pretty crude, and in time, we've become experts at outwitting them. In an ideal world (at least from the perspective of our genes), the parts of our brain that decide which activities are pleasurable would be extremely fussy, responding only to things that are truly good for us. For example, fruits have sugar, and mammals need sugar, so it makes sense that we should

---

*My friend Brad, who hates to see me suffer abstemiously in the service of some abstract long-term good, likes to bring me to a restaurant called Blue Ribbon Sushi, where he invariably orders the green tea crème brûlée. Usually, despite my best intentions, we wind up ordering two.

have evolved a "taste" for fruit; all well and good. But those sugar sensors can't tell the difference between a real fruit and a synthetic fruit that packages the flavor without the nutrition. We humans (collectively, if not as individuals) have figured out thousands of ways to *trick* our pleasure centers. Does the tongue like sweetness of fruit? Aha! Can I interest you in some Life Savers? Orange soda? Fruit juice made entirely from artificial flavors? A ripe watermelon may be good for us, but a watermelon-flavored candy is not.

And watermelon-flavored candy is only the start. The vast majority of the mental mechanisms we use for detecting pleasure are equally crude, and thus easily hoodwinked. In general, our pleasure detectors tend to respond not just to some specific stimulus that might have been desirable in the environment of our ancestors, but to a whole array of other stimuli that may do little for our genes. The machinery for making us enjoy sex, for example, causes us to revel in the activity, just as any reasonable evolutionary psychologist would anticipate, but not just when sex might lead to offspring (the narrowest tuning one might imagine), or even to pair bonding, but much more broadly: at just about any time, under almost any circumstance, in twos and threes and solo, with people of the same sex and with people of the opposite sex, with orifices that contribute to reproduction and with other body parts that don't. Every time a person has sex without directly or indirectly furthering their reproductivity, some genes have been fooled.

The final irony, of course, is that even though sex is incredibly motivating, people often have it in ways that are deliberately designed *not* to produce children. Heterosexuals get their tubes tied, gay men continue to have unprotected sex in the era of HIV, and pedophiles pursue their interests even when they risk prison and community censure. From the perspective of genes, all of this, aside from sex for reproduction or parental pair bonding, is a giant mistake.

To be sure, evolutionary psychologists have tried to find adaptive value in at least one of these variations (homosexuality), but none of the explanations are particularly compelling. (There is, for example, the "gay uncle" hypothesis, according to which homosexuality persists in the population because gay people often invest considerable resources in the offspring of their siblings.)* A more reasonable accounting, in my view, is that homosexuality is just like any other variation on sexuality, an instance of a pleasure system that was *only broadly tuned* (toward intimacy and contact) rather than *narrowly focused* (on procreation) by evolution, co-opted for a function other than the one to which it was strictly adapted. Through a mixture of genetics and experience, people can come to associate all manner of different things with pleasure, and proceed on that basis.†

The situation with sex is fairly typical. A substantial portion of our mental machinery seems to exist in order to assess reward (a proxy for pleasure), but virtually all of that machinery allows a broader range of options than might (from a gene's-eye view) be ideal. We see this with enjoyment of sugar — a hot fudge sundae just about always brings pleasure, whether we need the calories or not — but also with more modern compulsions, like addiction to the Internet. This compulsion presumably begins with an ancestral circuit that rewarded us for obtaining information. As the psychologist George Miller put it, we are all "informavores," and it's easy to see how ancestors who liked to gather facts might have outpropagated those who showed little interest in learning new things. But once

---

*The trouble is there's no evidence that all that good uncle-ing (for relatives that are only one-eighth genetically related) offsets the direct cost of failing to reproduce. Other popular adaptationist accounts of homosexuality include the Sneaky Male theory (favored by Richard Dawkins) and the Spare Uncle theory, by which an uncle who stays home from the hunt can fill in for a dad who doesn't make it home.

†If homosexuality is a sort of evolutionary byproduct, rather than a direct product of natural selection, does that make it wrong to be gay? Not at all; the morality of sexuality should depend on *consent,* not evolutionary origin. Race is biological, religion is not, but we protect both. By the same token, I see pedophilia as immoral — not because it is not procreative but simply because one party in the equation is not mature enough to genuinely give consent; likewise, of course, for bestiality.

again we have a system that hasn't been tuned precisely enough: it's one thing to get a kick from learning which herbs help cure open wounds, but another to get a kick from learning the latest on Angelina and Brad. We would probably all be better off if we were choosier about what information we sought, à la Sherlock Holmes, who notoriously didn't even know that the earth revolved around the sun. His theory, which we could perhaps learn from, goes like this:

> A man's brain originally is like a little empty attic, and you have to stock it with such furniture as you choose. A fool takes in all the lumber of every sort that he comes across, so that the knowledge which might be useful to him gets crowded out, or at best is jumbled up with a lot of other things, so that he has a difficulty in laying his hands upon it . . . It is of the highest importance, therefore, not to have useless facts elbowing out the useful ones.

Alas, Sherlock Holmes is only fictional. Few people in the real world have information-gathering systems as disciplined or as finely tuned as his. Instead, for most of us, just about *any* information can drive our pleasure meters. Late at night, if I allow myself access to the Internet, I'm liable to *click here* (World War II), click *there* (Iwo Jima), and then mindlessly follow *another link* (Clint Eastwood), only to stumble on a *fourth* (*Dirty Harry*), rapidly chaining my way from topic to topic, with no clear destination in mind. Yet each tidbit brings me pleasure. I'm not a history scholar, I'm not a film critic; its unlikely that any of this information will ever come in handy. But I can't help it; I just *like* trivia, and my brain isn't wired precisely enough to make me more discriminating. Want to stop my web surfing? Go ahead, make my day.

Something similar happens in our eternal quest for control. Study after study has shown that a sense of control makes people feel happy. One classic study, for example, put people in a position of listening to a series of sudden and unpredictable noises, played at excruciatingly random intervals. Some subjects were led to believe that they could do something about it — press a button to stop the noise — but others were told that they were powerless. The empowered subjects were less stressed and more happy — even though they hardly ever actually pressed the button. (Elevator "door close" buttons work on a similar principle.) Again, there would be adaptive sense in a narrowly focused system: organisms that sought out environments in which they had a measure of control would outcompete those that left themselves entirely at the mercy of stronger forces. (Better, for example, to wade in a slowly moving stream than a full-on waterfall.) But in modern life, we trick the machinery that rewards us for a sense of accomplishment, spending hours and hours perfecting golf swings or learning how to create a perfect piece of pottery — without discernibly increasing the number or quality of our offspring.

More generally, modern life is full of what evolutionary psychologists call "hypernormal stimuli," stimuli so "perfect" they don't exist in the ordinary world: the anatomically impossible measurements of Barbie, the airbrushed sheen of a model's face, the fast, sensation-filled jump cuts of MTV, and the artificial synthesized drum beats of the nightclub. Such stimuli deliver a purer kick than anything could in the ancestral world. Video games are a perfect case in point; we enjoy them because of the sense of control they afford; we like them to the extent that we can succeed in the challenges they pose — and we cease to enjoy them the minute we lose that sense of a control. A game that doesn't seem fair doesn't seem fun, precisely because it doesn't yield a sense of mastery. Each new level of challenge is designed to intensify that kick. Video games aren't just about control; they are the distillation of control: hypernormal variations on the naturally rewarding process of skill learning, designed to deliver as frequently as possible the kick associated with mastery. If video games (produced by an industry racking up billions of dollars in sales each year) strike some people as more fun than life itself, genes be damned, it is precisely because the games have been designed to exploit the intrinsic imprecision of our mechanism for detecting pleasure.

In the final analysis, pleasure is an eclectic thing. We love information, physical contact, social contact, good food, fine wine, time with our pets, music, theater, dancing, fiction, skiing, skateboarding, and video games; sometimes we pay people money to get us drunk and make us laugh. The list is virtually endless. Some evolutionary psychologists have tried to ascribe adaptive benefits to many of these phenomena, as in Geoffrey Miller's suggestion that music evolved for the purpose of courtship. (Another popular hypothesis is that music evolved for the purpose of singing lullabies.) Miller's flagship example is Jimi Hendrix:

> This rock guitarist extraordinaire died at the age of 27 in 1970, overdosing on the drugs he used to fire his musical imagination. His music output, three studio albums and hundreds of live concerts, did him no survival favours. But he did have sexual liaisons with hundreds of groupies, maintained parallel long-term relationships with at least two women, and fathered at least three children in the U.S., Germany, and Sweden. Under ancestral conditions before birth control, he would have fathered many more.

But none of these hypotheses is especially convincing. The sexual selection theory, for instance, predicts that males ought to have more musical talent than females, but even if teenage boys have been known to spend untold hours jamming in pursuit of the world's heaviest metal, there's no compelling evidence that males actually have greater musical talent.* There are thousands (or perhaps hundreds of thousands) of happily married women who devote their lives

---

*All this is with respect to humans. The bird world is a different story; there, males do most of the singing, and the connection to courtship is more direct.

to playing, composing, and recording music. What's more, there's no particular reason to think that the alleged seducees (women, in Miller's account) derive any less pleasure from making music than do the alleged seducers, or that an appreciation of music is in any way tied to fertility. No doubt music *can* be used in the service of courtship, but the fact that a trait can be used in a particular way doesn't prove that it evolved *for* that purpose; likewise, of course, with lullabies.

Instead, many modern pleasures may emerge from the broadly tuned pleasure systems that we inherited from our ancestors. Although music as such — used for purposes of recreation, not mere identification (the way songbirds and cetaceans employ musical sounds) — is unique to humans, many or most of the cognitive mechanisms that underlie music are not. Just as much of language is built on brain circuits that are considerably ancient, there is good reason to think that music relies largely (though perhaps not entirely) on devices that we inherited from our premusical ancestors. Rhythmic production appears in rudimentary form in at least some apes (King Kong isn't alone in beating his chest), and the ability to differentiate pitch is even more widespread. Goldfish and pigeons have been trained to distinguish musical styles. Music likely also taps into the sort of pleasure we (and most apes) derive from social intimacy, the enjoyment we get from accurate predictions (as in the anticipation of rhythmic timing) and their juxtaposition with the unexpected,* and something rather more mundane, the "mere familiarity

---

*Music that is either purely predictable or completely unpredictable is generally considered unpleasant — tedious when it's too predictable, discordant when it's too unpredictable. Composers like John Cage have, of course played, with that balance, but few people derive the same pleasure from Cage's quasi-random ("aleatoric") compositions that they do from music with a more traditional balance between the predictable and the surprising — a fact that holds true in genres ranging from classical to jazz and rock. (The art of improvisation is to invent what in hindsight seems surprising yet inevitable.)

effect" (mentioned earlier, in the context of belief). And in playing musical instruments (and in singing), we get a sense of mastery and control. When we listen to the blues, we do so, at least in part, so we won't feel alone; even the most angst-ridden teenager gets some pleasure in knowing that his or her pain is shared.

Forms of entertainment like music, movies, and video games might be thought of as what Steven Pinker calls "pleasure technologies" — cultural inventions that maximize the responses of our reward system. We enjoy such things not because they propagate our genes or because they conveyed specific advantages to our ancestors, but because they have been *culturally* selected for — precisely to the extent that they manage to tap into loopholes in our preexisting pleasure-seeking machinery.

The bottom line is this: our pleasure center consists not of some set of mechanisms perfectly tuned to promote the survival of the species, but a grab bag of crude mechanisms that are easily (and pleasurably) outwitted. Pleasure is only loosely correlated with what evolutionary biologists call "reproductive fitness" — and for that, we should be grateful.

Given how much we do to orient ourselves to the pursuit of pleasure, you'd expect us to be pretty good at assessing what's likely to make us happy and what's not. Here again, evolution holds some surprises.

A simple problem is that much of what makes us happy doesn't last long. Candy bars make us happy — for an instant — but we soon return to the state of mind we experienced before we had one. The same holds (or can hold) for sex, for movies, for television shows, and for rock concerts. Many of our most intense pleasures are shortlived.

But there's a deeper issue, which shows up in how we set our long-term goals; although we behave as if we want to maximize our long-term happiness, we frequently are remarkably poor at anticipating what will genuinely make us happy. As the psychologists Timothy Wilson and Daniel Gilbert have shown, predicting our own happiness can be a bit like forecasting the weather: a pretty inexact science. Their textbook case,* which should give pause to assistant professors everywhere, concerns the young faculty member's inevitable quest for tenure. Virtually every major U.S. institution promises to its finest, most successful young professors a lifetime of academic freedom and guaranteed employment. Slog through graduate school, a postdoc or two, and five or six years of defining your very own academic niche, and if you succeed (as measured by the length of your résumé), you will gain tenure and be set for life.

The flip slide (rarely mentioned) is the slog that fails. Five to ten years spent working on a Ph.D., the postdocs, the half-decade of teaching, unappreciative undergraduates, the interminable faculty meetings, the struggle for grant money — and for what? Without a publication record, you're out of a job. Any professor can tell you that tenure is fantastic, and not getting tenure is miserable.

Or so we believe. In reality, neither outcome makes nearly as much difference to overall happiness as people generally assume. People who get tenure tend to be relieved, and initially ecstatic, but their happiness doesn't linger; they soon move on to worrying about other things. By the same token, people who don't get tenure are indeed often initially miserable, but their misery is usually short-lived. Instead, after the initial shock, people generally adapt to their circumstances. Some realize that the academic rat race isn't for them; others start new careers that they actually enjoy more.

---

*Gilbert has another favorite example: children. Although most people anticipate that having children will increase their net happiness, studies show that people with children are actually less happy on average than those without. Although the highs ("Daddy I wuv you") may be spectacular, on a moment-by-moment basis, most of the time spent taking care of children is just plain work. "Objective" studies that ask people to rate how happy they are at random moments rank raising children — a task with clear adaptive advantage — somewhere between housework and television, well below sex and movies. Luckily, from the perspective of perpetuating the species, people tend to remember the intermittent high points better than the daily grind of diapers and chauffeur duty.

Aspiring assistant professors who think that their future happiness hinges on getting tenure often fail to take into account one of the most deeply hard-wired properties of the mind: the tendency to get used to whatever's going on. The technical term for that is *adaptation*.* For example, the sound of rumbling trucks outside your office may annoy you at first, but over time you learn to block it out — that's adaptation. Similarly, we can adapt to even more serious annoyances, especially those that are predictable, which is why a boss who acts like a jerk every day can actually be less irritating than a boss who acts like a jerk less often, but at random intervals. As long as something is a constant, we can learn to live with it. Our circumstances do matter, but psychological adaptation means that they often matter less than we might expect.

This is true at both ends of the spectrum. Lottery winners get used to their newfound wealth, and others, people like the late Christopher Reeve, find ways of coping with circumstances most of us would find unimaginable. Don't get me wrong — I'd like to win the lottery and hope that I will never be seriously injured. But as a psychologist I know that winning the lottery wouldn't really change my life. Not only would I have to fend off all the long-lost "friends" who would come out of the woodwork, but also I'd face the inevitable fact of adaptation: the initial rush couldn't last because the brain won't allow it to.

The power of adaptation is one reason why money matters a lot less than most people think. According to literary legend, F. Scott Fitzgerald once said to Ernest Hemingway, "The rich are not like us." Hemingway allegedly brushed him off with the reply "Yes, they have more money," implying that wealth alone might make little difference. Hemingway was right. People above the poverty line are

---

*The psychological use of the term is, of course, distinct from the evolutionary use. In psychology, *adaptation* refers to the process of becoming accustomed to something such that it becomes familiar; in evolution, it refers to a trait that is selected over the space of evolutionary time.

happier than people below the poverty line, but the truly wealthy aren't that much happier than the merely rich. One recent study, for example, showed that people making over $90,000 a year were no happier than those in the $50,000–$89,999 bracket. A recent *New York Times* article described a support group for multimillionaires. Another study reported that although average family income in Japan increased by a factor of five from 1958 to 1987, people's self-reports of happiness didn't change at all; all that extra income, but no extra happiness. Similar increases in the standard of living have occurred in the United States, again, with little effect on overall happiness. Study after study has shown that wealth predicts happiness only to a small degree. New material goods often bring tremendous initial pleasure, but we soon get used to them; that new Audi may be a blast to drive at first, but like any other vehicle, eventually it's just transportation.

Ironically, what really seems to matter is not absolute wealth, but relative income. Most people would rather make $70,000 at a job where their co-workers average $60,000 than $80,000 at a job where co-workers average $90,000. As a community's overall wealth increases, individual expectations expand; we don't just want to be rich, we want to be *richer* (than our neighbors). The net result is that many of us seem to be on a happiness treadmill, working harder and harder to maintain essentially the same level of happiness.

One of the most surprising things about happiness is just how poor we are at measuring it. It's not just that no brain scanner or dopamine counter can do a good job, but that we often just don't know — yet another hint of how klugey the whole apparatus of happiness really is.

Are you happy right now, at this very moment, reading this very book? Seriously, how would you rate the experience, on a scale of 1 ("I'd rather being doing the dishes") to 7 ("If this were any more fun, it would be illegal!")? You probably feel that you just "know" or can "intuit" the answer — that you can directly assess how happy you are, in the same way that you can determine whether you're too hot or too cold. But a number of studies suggest that our impression of direct intuition is an illusion.

Think back to that study of undergraduates who answered the question "How happy are you?" after first recounting their recent dating history. We're no different from them. Asking people about their overall happiness just after inquiring into the state of their marriage or their health has a similar effect. These studies tell us that people often don't really know how happy they are. Our subjective sense of happiness is, like so many of our beliefs, fluid, and greatly dependent on context.

Perhaps for that reason, the more we think about how we happy we are, the less happy we become. People who ruminate less upon their own circumstances tend to be happier than those who think about them more, just as Woody Allen implied in *Annie Hall*. When two attractive yet vacant-looking pedestrians walk by, Allen's character asks them to reveal the secret of their happiness. The woman answers first: "I'm very shallow and empty and I have no ideas and nothing interesting to say," to which her handsome boyfriend adds, "And I'm exactly the same way." The two stride gaily away. In other words, to paraphrase Mark Twain, dissecting our own happiness may be like dissecting frogs: both tend to die in the process.

Our lack of self-understanding may seem startling at first, but in hindsight it should scarcely seem surprising. Evolution doesn't "care" whether we understand our own internal operations, or even whether we are happy. Happiness, or more properly, the opportunity to pursue it, is little more than a motor that moves us. The happiness treadmill keeps us going: alive, reproducing, taking care of children, surviving for another day. Evolution didn't evolve us to *be* happy, it evolved us to *pursue* happiness.

In the battle between us and our genes, the kicker is this: to the extent that we see pleasure as a compass (albeit a flawed one) that tells us where we should be headed, and to the extent that we see happiness as a thermometer that tells us how we are doing, those instruments should, by rights, *be instruments we can't fool with.* Had our brains been built from scratch, the instruments that evaluated our mental state would no doubt behave a little like the meters electric companies use, which are instruments that we can inspect but not tinker with. No sensible person would buy a thermometer that displayed only the temperature that its owner wanted, rather than the actual temperature. But humans routinely try to outwit their instruments. Not just by seeking new ways of getting pleasure, but by lying to ourselves when we don't like what our happy-o-meter tells us. We "acquire" tastes (in an effort to override our pleasure compass), and, more significantly, when things aren't going well, we try to persuade ourselves that everything is fine. (We do the same thing with pain, every time we pop an Advil or an aspirin.)

Take, for example, your average undergraduate around the time I hand out grades. Students who get A's are thrilled, they're happy, and they accept their grades with pleasure, even glee. People with C's are, as you might imagine, less enthusiastic, dwelling for the most part not on what *they* did wrong, but what *I* did wrong. (Question 27 on the exam wasn't fair, we never talked about that in class, and how could Professor Marcus have taken off three points for my answer to question 42?) Meanwhile, it never occurs to the keeners in the front that I might have mistakenly been *too* generous to them. This asymmetry reeks, of course, of motivated reasoning, but I don't mean to complain; I do the same thing, ranting and raving at reviewers who reject my papers, blessing (rather than questioning) the wisdom of those who accept them. Similarly, car accidents are never our fault — it's always the other guy.

Freud would have seen all this self-deception as an illustration of what he called "defense mechanisms"; I see it as motivated reasoning. Either way, examples like these exemplify our habit of trying to fool the thermometer. Why feel bad that we've done something wrong when we can so easily jiggle the thermometer? As Jeff Goldblum's character put it in *The Big Chill,* "Rationalizations are more important than sex." "Ever go a week without a rationalization?" he asked.

We do our best to succeed, but if at first we don't succeed, we can always lie, dissemble, or rationalize. In keeping with this idea, most Westerners believe themselves to be smarter, fairer, more considerate, more dependable, and more creative than average. And — shades of Garrison Keillor's Lake Wobegon, where "the women are strong, the men are good looking, and all the children are above average" — we also convince ourselves that we are better-than-average drivers and have better-than-average health prospects. But you do the math: we can't *all* be above average. When Muhammad Ali said "I'm the greatest," he spoke the truth; the rest of us are probably just kidding ourselves (or at least our happy-o-meters).

Classic studies of a phenomenon called "cognitive dissonance" make the point in a different way.* Back in the late 1950s, Leon Festinger did a famous series of experiments in which he asked subjects (undergraduate students) to do tedious menial tasks (such as sticking a set of plain pegs into an plain board). Here's the rub: some subjects were paid well ($20, a lot of money in 1959), but others, poorly ($1). Afterward, all were asked how much they liked the task. People who were paid well typically confessed to being bored, but people who were paid only a dollar tended to delude themselves into thinking that putting all those pegs into little holes was fun. Evidently they didn't want to admit to themselves that they'd wasted their time. Once again, who's directing whom? Is happiness guiding

---

*The term *cognitive dissonance* has crossed over into popular culture, but its proper meaning hasn't. People use it informally to refer to any situation that's disturbing or unexpected. ("Dude, when he finds out we crashed his mother's car, he's going to be feeling some *major* cognitive dissonance.") The original use of the term refers to something less obvious, but far more interesting: the tension we feel when we realize (however dimly) that two or more of our beliefs are in conflict.

us, or are we micromanaging our own guide? It's as if we paid a sherpa to guide us up a mountain — only to ignore him whenever he told us we were going in the wrong direction. In short, we do everything in our power to make ourselves happy and comfortable with the world, but we stand perfectly ready to lie to ourselves if the truth doesn't cooperate.

Our tendency toward self-deception can lead us to lie not just about ourselves but about others. The psychologist Melvyn Lerner, for example, identified what he called a "Belief in a Just World"; it feels better to live in a world that seems just than one that seems unjust. Taken to its extreme, that belief can lead people to do things that are downright deplorable, such as blaming innocent victims. Rape victims, for example, are sometimes perceived as if they are to blame, or "had it coming." Perhaps the apotheosis of this sort of behavior occurred during the Irish potato famine, when a rather objectionable English politician said that "the great evil with which we have to contend is not the physical evil of the famine, but the moral evil of the selfish, perverse, and turbulent character of the people." Blaming victims may allow us to cling to the happy notion that the world is just, but its moral costs are often considerable.

A robot that was more sensibly engineered might retain the capacity for deliberative reason but dispense with all the rationalization and self-deception. Such a robot would be aware of its present state but prepared, Buddha-like, to accept it, good or bad, with equanimity rather than agony, and thus choose to take actions based on reality rather than delusion.

In biological terms, the neurotransmitters that underlie emotion, such as dopamine and serotonin, are ancient, tracing their history at least to the first vertebrates and playing a pivotal role in the reflexive systems of animals including fish, birds, and even mammals. Humans, with our massive prefrontal cortex, add substantial reflective reasoning on top, and thus we find ourselves with an instrument-fooling kluge. Virtually every study of reasoned decision making locates this capacity in the prefrontal cortex; emotion is attributed to the limbic system (and oribitofrontal cortex). A spot known as the anterior cingulate, souped up in human beings and other great apes, seems to mediate between the two. Deliberative prefrontal thought is piled *on top of* automatic emotional feelings — it doesn't replace them. So we've wound up with a double-edged kluge: our id perpetually at war with our ego, short-term and long-term desires never at peace.

What's the best evidence for this split? Teenagers. Teenagers as a species seem almost pathologically driven by short-term rewards. They make unrealistic estimates of the attendant risks and pay little attention to long-term costs. Why? According to one recent study, the nucleus accumbens, which assesses reward, matures before the orbital frontal cortex, which guides long-term planning and deliberative reasoning. Thus teenagers may have an adult capacity to appreciate short-term gain, but only a child's capacity to recognize long-term risk.

Here again, evolutionary inertia takes precedence over sensible design. Ideally, our judicious system and our reflexive system would mature at comparable rates. But perhaps because of the dynamics of how genomes change, biology tends, on average, to put together the evolutionarily old before the evolutionarily new. The spine, for example, a structure that is shared by all vertebrates, develops before the toes, which evolved more recently. The same thing happens with the brain — the ancestral precedes the modern, which perhaps helps us understand why teens, almost literally, often don't know what to do with themselves. Pleasure, in the context of a system not yet fully wired, can be a dangerous thing. Pleasure giveth, and pleasure taketh away.

7
THINGS FALL APART

> I can calculate the motion of heavenly bodies but not the madness of people.
>
> — SIR ISAAC NEWTON

ENGINEERS WOULD PROBABLY build kluges more often if it were not for one small fact: that which is clumsy is rarely reliable. Kluges are often (though not always) designed to last for a moment, not a lifetime. On *Apollo 13*, with time running out and the nearest factory 200,000 miles away, making a kluge was essential. But the fact that some clever NASA engineers managed to build a substitute air-filter adapter using duct tape and a sock doesn't mean that what they built was well built; the whole thing could have fallen apart at a moment's notice. Even kluges designed to last for a while — like vacuumed-powered windshield wipers — often have what engineers might call "narrow operating conditions." (You wanted those wipers to work *uphill* too?)

There can be little doubt that the human brain too is fragile, and not just because it routinely commits the cognitive errors we've already discussed, but also because it is deeply vulnerable both to minor malfunctions and even, in some cases, severe breakdown. The mildest malfunctions are what chess masters call blunders and a Norwegian friend of mine calls "brain farts" — momentary lapses of reason and attention that cause chagrin (d'Oh!) and the occasional traffic accident. We know better, but for a moment we just plain goof. Despite our best intentions, our brain just doesn't manage to do what we want it to. No one is immune to this. Even Tiger Woods occasionally misses an easy putt.

At the risk of stating the obvious, properly programmed computers simply don't make these kinds of transient blunders. My laptop has never, ever forgotten to "carry the one" in the midst of a complicated sum, nor (to my chagrin) has it "spaced out" and neglected to protect its queen during a game of chess. Eskimos don't really have 500 words for snow, but we English speakers sure have a lot of words for our cognitive short circuits: not just *mistakes, blunders,* and *fingerfehlers* (a hybrid of English and German that's popular among chess masters) but also *goofs, gaffes, flubs,* and *boo-boos,* along with *slips, howlers, oversights,* and *lapses.* Needless to say, we have plenty of opportunities to use this vocabulary.

The fact that even the best of us are prone to the occasional blunder illustrates something important about the neural hardware that runs our mental software: consistency just isn't our forte. Nearly everything we carbon-based units do runs some chance of error. Word-finding failures, moments of disorientation and forgetfulness, each in its own way points to the imperfection inherent in the nerve cells (neurons) from which brain circuits are made. If a foolish consistency is the hobgoblin of little minds, as the American writer Ralph Waldo Emerson (1803–1882) once said, a foolish *inconsistency* characterizes every single human mind. There's no guarantee that any person's mind will always fire on all cylinders.

Yet random gaffes and transient blunders are just a tiny piece of a larger, more serious puzzle: why do we humans so often fail to do what we set out to do, and what makes the mind so fragile that it can sometimes spiral out of control altogether?

Plenty of circumstances systematically increase the chance of making mental errors. The more that's on our mind, for example, the more likely we are to fall back on our primitive ancestral system. Bye-bye, prefrontal cortex, signature of the noble human mind; hello, animal instinct, short-sighted and reactive. People committed to eating in a healthful way are, for example, more likely to turn to junk if something else is on their mind. Laboratory studies show that as the demands on the brain, so-called *cognitive load,* increase, the ancestral system continues business as usual — while the more modern deliberative system gets left behind. Precisely when the cognitive chips are down, when we most need our more evolved (and theoretically sounder) faculties, they can let us down and leave us *less* judicious. When mentally (or emotionally) taxed, we become more prone to stereotyping, more egocentric, and more vulnerable to the pernicious effects of anchoring.

No system, of course, can cope with infinite demands, but if I had been hired to design this aspect of the mind, I would have started by letting the deliberative, "rational" system take priority, whenever time permits, favoring the rational over the reflexive where possible. In giving precedence instead to the ancestral reflexive system — not necessarily because it is better but simply because it is older — evolution has squandered some of our most valuable intellectual resources.

Whether we are under cognitive strain or not, another banal but systematic failure hampers our ability to meet mental goals: most of us — at one time or another — "space out." We have one thing that we nominally intend to accomplish (say, finishing a report before a deadline), and the next thing you know, our thoughts have wandered. An ideal creature would be endowed with an iron will, sticking, in all but the most serious emergencies, to carefully constructed goals. Humans, by contrast, are characteristically distractible, no matter what the task might be.

Even with the aid of Google, I can neither confirm nor deny the widespread rumor that one in four people is daydreaming about sex at any given moment,* but my hunch is that the number is not too far from the truth. According to a recent British survey, during office

---

*Another study, commissioned by a soap manufacturer, suggests that in the shower "men split their time daydreaming about sex (57 percent) and thinking about work (57 percent)." As Dave Barry put it on his blog, "This tells us two things: (1) Men lie to survey-takers. (2) Survey-takers do not always have a solid understanding of mathematics."

meetings one in three office workers reportedly daydreams about sex. An economist quoted in the UK's *Sunday Daily Times* estimates that this daydreaming may cost the British economy about £7.8 billion annually.

If you're not the boss, statistics on daydreaming about sex might be amusing, but "zoning out," as it is known in the technical literature, is a real problem. For example, all told, nearly a 100,000 Americans a year die in accidents of various sorts (in motor vehicles or otherwise); if even a third of those tragedies are due to lapses of attention, mind wandering is one of the top ten leading causes of death.*

My computer never zones out while downloading my email, but I find my mind wandering all the time, and not just during faculty meetings; to my chagrin, this also happens during those rare moments when I have time for pleasure reading. Attention-deficit disorder (ADD) gets all the headlines, but in reality, nearly *everyone* periodically finds it hard to stay on task.

What explains our species-wide tendency to zone out — even, sometimes, in the midst of important things? My guess is that our inherent distractibility is one more consequence of the sloppy integration between an ancestral, reflexive set of goal-setting mechanisms (perhaps shared with all mammals) and our evolutionarily more recent deliberative system, which, clever as it may be, aren't always kept in the loop.

Even when we aren't zoning out, we are often chickening out: putting off till tomorrow what we really ought to do today. As the eighteenth-century lexicographer and essayist Samuel Johnson put it (some 200 years before the invention of video games), procrastination is "one of the general weaknesses, which in spite of the instruction of moralists,

---

*One recent NHTSA study suggests that fully 80 percent of fender-benders can be attributed to inattention. Among fatal car accidents, no firm numbers are available, but we know that about 40 percent are attributable to alcohol; among the remaining 60 percent involving sober drivers, inattention likely plays a major role.

and the remonstrances of reason, prevail to a greater or less degree in every mind."

By one recent estimate, 80–95 percent of college students engage in procrastination, and two thirds of all students consider themselves to be (habitual) procrastinators. Another estimate says that 15–20 percent of all adults are chronically affected — and I can't help but wonder whether the rest are simply lying. Most people are troubled by procrastination; most characterize it as bad, harmful, and foolish. And most of us do it anyway.

It's hard to see how procrastination per se could be adaptive. The costs are often considerable, the benefits minuscule, and it wastes all the mental effort people put into making plans in the first place. Studies have shown that students who routinely procrastinate consistently get lower grades; businesses that miss deadlines due to the procrastination of their employees can sometimes lose millions of dollars. Yet many of us can't help ourselves. Why, when so little good comes of procrastinating, do we persist in doing it so much?

I for one hope someone figures out the answer, and soon, maybe even inventing a magic pill that can keep us on task. Too bad no one's gotten around to it just yet: tomorrow and tomorrow and tomorrow. In the meantime, the research that *has* been done suggests a diagnosis if not a solution: procrastination is, in words of one psychologist, the "quintessential self-regulatory failure." Nobody, of course, can at a given moment do all of the things that need to be done, but the essence of procrastination is the way in which we defer progress on our own most important goals.

The problem, of course, is not that we put things off, per se; if we need to buy groceries and do our taxes, we literally can't accomplish both at the same time. If we do one now, the other must wait. The problem is that we often postpone the things that *need* to get done in favor of others — watching television or playing video games — that most decidedly don't need to get done. Procrastination is a sign of our inner kluge because it shows how our top-level goals (spend more time with the children, finish that novel) are routinely undermined by goals of considerably less priority (if catching up on the latest episodes of *Desperate Housewives* can be counted as a "goal" at all).

People need downtime and I don't begrudge them that, but procrastination does highlight a fundamental glitch in our cognitive "design": the gap between the machinery that sets our goals (offline) and the machinery that chooses (online, in the moment) which goals to follow.

The tasks most likely to tempt us to procrastinate generally meet two conditions: we don't enjoy doing them and we don't have to do them *now*. Given half a chance, we put off the aversive and savor the fun, often without considering the ultimate costs. Procrastination is, in short, the bastard child of future discounting (that tendency to devalue the future in relation to the present) and the use of pleasure as a quick-and-dirty compass.

We zone out, we chicken out, we deceive. To be human is to fight a lifelong uphill battle for self-control. Why? Because evolution left us clever enough to set reasonable goals but without the willpower to see them through.

Alas, zoning out and chickening out are among the least of our problems; the most serious are the psychological breakdowns that require professional help. From schizophrenia to obsessive-compulsive disorder and bipolar disorder (also called manic depression), nothing more clearly illustrates the vulnerability of the human mind than our susceptibility to chronic and severe mental disorders. What explains the madness of John Nash, the bipolar disorder of Vincent van Gogh and Virginia Woolf, the paranoia of Edgar Allan Poe, the obsessive-compulsive disorder of Howard Hughes, the depression that drove Ernest Hemingway, Jerzy Kosinski, Sylvia Plath, and Spalding Gray to suicide? Perhaps a quarter of all human beings at a given moment suffer from one clinical disorder or another. And, over the course of a lifetime, almost half the population will face bouts of one mental illness or another. Why is our mind so prone to breakdown?

Let's start with a fact that is well known but perhaps not fully appreciated. For the most part, mental disorders aren't random unprecedented anomalies, completely unique to the individuals who suffer from them. Rather, they comprise clusters of symptoms that recur again and again. When things fall apart mentally, they tend to do so in recognizable ways, what engineers sometimes call "known failure modes." A given make and model of a car, say, might have a fine engine but consistently suffer from electrical problems. The human mind is vulnerable to its own particular malfunctions, well documented enough to be classified in the human equivalent of Chilton's *Auto Repair*: the *DSM-IV* (short for *Diagnostic and Statistical Manual of Mental Disorders*, fourth edition; a fifth edition is scheduled for 2011).

To be sure, symptoms vary among individuals, both in severity and in number. Just as no two colds are exactly alike, no two people diagnosed with a given mental illness experience it in precisely the same way. Some people with depression, for example, are dysfunctional, and some aren't; some people with schizophrenia hear voices, and others don't.

And diagnosis remains an inexact science. There are a few disorders (such as multiple personality syndrome) whose very existence is controversial, and a few "conditions" used to be labeled as disorders but never should have been (such as homosexuality, removed from the *DSM-III* in 1973).* But by and large, there is an astonishing amount of consistency in the ways in which the human mind can break down, and certain symptoms, such as dysphoria (sadness), anxiety, panic, paranoia, delusions, obsessions, and unchecked aggression, recur again and again.

When we see the same basic patterns over and over, there has to

---

*In an earlier era, there was "the inability to achieve vaginal orgasm," "childhood masturbation disorder," and drapetomania, the inexplicable desire on the part of some slaves to run away, or what I like to think of as freedom sickness.

be a reason for them. What is the mind, such that it breaks down in the ways that it does?

The standard tack in evolutionary psychiatry, the branch of evolutionary psychology that deals with mental disorders, is to explain particular disorders (or occasionally symptoms) in terms of hidden benefits.* We saw one example in the first chapter, the somewhat dubious suggestion that schizophrenia might have been selected for by natural selection because of a purported benefit that visions conveyed to tribal shamans, but there are many others. Agoraphobia has been viewed as a "potentially adaptive consequence of repeated panic attacks," and anxiety has been interpreted as a way of "altering our thinking, behavior, and physiology in advantageous ways." Depression, meanwhile, allegedly evolved as a way of allowing individuals to "accept defeat . . . and accommodate to what would otherwise be unacceptably low social rank."

If you're like me, you won't find these examples particularly compelling. Were schizophrenics really more likely than other people to become shamans? Were those who became shamans more successful than their non-schizophrenic counterparts in producing viable offspring? Even if they were, are shamans prevalent enough in history to explain why at least 1 in every 100 humans suffers from the disorder? The depression theory initially seems more promising; as the au-

---

*Another strand of evolutionary psychology emphasizes the extent to which current environments differ from those of our ancestors. As Kurt Vonnegut Jr. (who before he became a novelist studied anthropology) put it, "It's obvious through the human experience that extended families and tribes are terribly important. We can do without an extended family as human beings about as easily as we do without vitamins or essential minerals." "Human beings will be happier," he wrote, "not when they cure cancer or get to Mars or eliminate racial prejudice or flush Lake Erie, but when they find ways to inhabit primitive communities again." I have great sympathy with this notion, but the stress of modern life is only a part of the story; as far as we can tell, mental disorders have been around for as long as humans have. Like virtually every aspect of mental life, mental illness depends on a mix of factors, some environmental, some biological.

thors note, it might well be better for the low man on the totem pole to accede to the wishes of an alpha male than to fight a battle that can't be won. Furthermore, depression often does stem from people's sense that their status is lower, relative to some peer group. But does the rest of the social competition theory even fit the facts? Depression isn't usually about *accepting* defeat, it's about *not* accepting it. A friend of mine, we'll call him T., has been clinically depressed for years. He's not particularly low in social rank (he's actually a man of considerable accomplishment). Yet although there is nothing objectively wrong with his life, he doesn't *accept* it: he ruminates on it. Depression hasn't mobilized him to improve his life, nor to keep him out of trouble; instead, it's paralyzed him, and it's difficult to see how paralysis could be adaptive.

Of course, I don't mean to suggest that one dubious theory is enough to rule out an entire line of work; certainly some physical disorders convey benefits, and there may well be analogous cases of mental disorders. The classic example of a physical disorder with a clear corresponding benefit is the gene that is associated with sickle cell anemia. Having two copies of the gene is harmful, but having a single copy of the gene alongside a normal copy can significantly reduce one's chance of contracting malaria. In environments where malaria has been widespread (such as sub-Saharan Africa), the benefits outweigh the potential costs. And, accordingly, copies of such genes are far more widespread among people whose ancestors lived in parts of the world where malaria was prevalent.

But while some physical disorders do demonstrably bring about offsetting benefits, most probably don't, and, with the possible exception of sociopathy,* I don't think I've ever seen a case of mental ill-

---

*Although sociopathy would be unlikely to spread through an entire population, it is not implausible that in a society in which *most* people were cooperative and trusting, a small number of sociopaths might survive and even thrive. Then again, at least in contemporary society, a fair number of sociopaths ultimately get caught and wind up in prison, with no further chance to reproduce and little opportunity to take care of offspring they might already have.

ness offering advantages that might convincingly outweigh the costs. There are few, if any, concrete illustrations of offsetting advantages in mental illness, no mental sickle-cell anemia that demonstrably protects again "mental malaria." Depression, for example, doesn't ward off anxiety (in the way that a propensity for sickling protects from malaria) — it co-occurs with it. Most of the literature on the alleged virtues of mental disorders simply seems fanciful. All too often, I am reminded of Voltaire's Dr. Pangloss, who found adaptive virtue in everything: "Observe, for instance, the nose is formed for spectacles, therefore we wear spectacles. The legs are visibly *designed* for stockings, accordingly we wear stockings. Stones were made to be hewn and to construct castles."

It's true that many disorders have at least some compensation, but the reasoning is often backward. The fact that some disorders have some redeeming features doesn't mean that those features *offset* the costs, nor does it necessarily explain why those disorders evolved in the first place. What happy person would volunteer to take a hypothetical depressant — call it "anti-Prozac" or "inverse Zoloft" — in order to accrue the benefits that allegedly accompany depression?

At the very least, it seems plausible that some disorders (or symptoms) may appear not as direct adaptations, but simply from inadequate "design" or outright failure. Just as cars run out of gas, the brain can run out of (or run low on) neurotransmitters (or the molecules that traffic in them). We are born with coping mechanisms (or the capacity to acquire them), but nothing guarantees that those coping mechanisms will be all powerful or infallible. A bridge that can withstand winds of 100 miles per hour but not 200 doesn't collapse in gusts of 200 miles per hour because it is *adaptive* to fail in such strong winds; it falls apart because it was built to a lesser specification. Similarly, other disorders, especially those that are extremely rare, may result from little more than "genetic noise," random mutations that convey no advantage whatsoever.

Even if we set aside possibilities like sheer genetic noise, it is a fallacy to assume that if a mental illness persists in a population, it must convey an advantage. The bitter reality is that evolution doesn't "care" about our inner lives, only results. So long as people with disorders reproduce at reasonably high rates, deleterious genetic variants can and do persist in the species, without regard to the fact that they leave their bearers in considerable emotional pain.*

All this has been discussed in the professional literature, but another possibility has gotten almost no attention: could it be that some aspects of mental illness persist not because of any specific advantage, but simply because evolution couldn't readily build us in any other way?

Take, for example, anxiety. An evolutionary psychologist might tell you that anxiety is like pain: both exist to motivate their bearers into certain kinds of action. Maybe so, but does that mean that anxiety is an *inevitable* component of motivation, which we would expect to see in any well-functioning organism? Not at all — anxiety might have goaded some of our prelinguistic, pre-deliberative-reasoning ancestors into action, but that doesn't make it the right system for creatures like us, who *do* have the capacity to reason. Instead, if we humans were built from the ground up, anxiety might have no place at all: our higher-level reasoning capacities could handle planning by themselves. In a creature empowered to set and follow its own goals, it's not clear that anxiety would serve any useful function.

One could make a similar argument about the human need for self-esteem, social approval, and rank — collectively, the source of much psychological distress. Perhaps in any world we could imagine, it would be to most creatures' benefit to secure social approval, but it is not clear why a lack of social approval ought necessarily to result in emotional pain. Why not be like the Buddhist robots I conjured in

---

*Evolution is also flagrantly unconcerned with the lives of those past child-bearing age; genes that predispose people to Huntington's chorea or Alzheimer's disease could bear some hidden benefit, but the disorder could persist even if it didn't, simply because, as something that happens late in life, the bottom line of reproductive fitness isn't affected.

the last chapter, always aware of (and responsive to) circumstances, but never troubled by them?

Science fiction? Who knows. What these thought experiments do tell us is that it is possible to imagine other ways in which creatures might live and breathe, and it's not clear that the disorders we see would inevitably evolve in those creatures.

What I am hinting at, of course, is this: the possibility that mental illness might stem, at least in part, from accidents of our evolutionary history. Consider, for example, our species-wide vulnerability to addiction, be it to cigarettes, alcohol, cocaine, sex, gambling, video games, chat rooms, or the Internet. Addiction can arise when short-term benefits appear subjectively enormous (as with heroin, often described as being better than sex), when long-term benefits appear subjectively small (to people otherwise depressed, who see themselves as having little to live for), or when the brain fails to properly compute the ratio between the two. (The latter seems to happen in some patients with lesions in the ventromedial prefrontal cortex, who evidently can detect costs and benefits but seem indifferent as to their ratio.) In each case, addiction can be thought of as a particular case of a general problem: our species-wide difficulty in balancing ancestral and modern systems of self-control.

To be sure, other factors are at work, such as the amount of pleasure a given individual gets from a given activity; some people get a kick out of gambling, and others would rather just save their pennies. Different people are vulnerable to different addictions, and to different degrees. But we are all at least somewhat at risk. Once the balance between long-term and short was left to a rather unprincipled tug-of-war, humanity's vulnerability to addiction may have become all but inevitable.

If the split in our systems of self-control represents one kind of fault line in the human mind, confirmation bias and motivated reasoning combine to form another: the relative ease with which humans can lose touch with reality. When we "lose it" or "blow things out of proportion," we lose perspective, getting so angry, for example, that all traces of objectivity vanish. It's not one of our virtues, but it is a part of being human; we are clearly a hotheaded species.

That said, most of the time, most of us get over it; we may lose touch in the course of an argument, but ultimately we take a deep breath or get a good night's sleep, and move on. ("Yes, it was really lousy of you to stay out all night and not call, but I admit that when I said you *never* call I might have been exaggerating. Slightly." Or, as Christine Lavin once sang, "I'm sorry, forgive me, . . . but I'm still mad at you.")

What occasionally allows normal people to spiral out of control is a witch's brew of cognitive kluges: (1) the clumsy apparatus of self-control (which in the heat of the moment all too often gives the upper hand to our reflexive system); (2) the lunacy of confirmation bias (which convinces us that we are *always* right, or nearly so); (3) its evil twin, motivated reasoning (which leads us to protect our beliefs, even those beliefs that are dubious); and (4) the contextually driven nature of memory (such that when we're angry at someone, we tend to remember other things about them that have made us angry in the past). In short, this leaves "hot" systems dominating cool reason; carnage often ensues.

That same mix, minus whatever inhibitory mechanisms normal people use to calm down, may exacerbate, or maybe even spawn, several other aspects of mental illness. Take, for example, the common symptom of paranoia. Once someone starts down that path — for whatever reason, legitimate or otherwise — the person may never leave it, because paranoia begets paranoia. As the old saying puts it, even the paranoid have real enemies; for an organism with confirmation bias and the will to deny counterevidence (that is, motivated reasoning), all that is necessary is one true enemy, if that. The paranoid person notices and recalls evidence that confirms his or her paranoia, discounts evidence that contradicts it, and the cycle repeats itself.

Depressives too often lose touch with reality, but in different ways. Depressives don't generally hallucinate (as, for example, many schizophrenics do), but they often distort their perception of reality by fixating on the negative aspects of their lives — losses, mistakes, missed opportunities, and so forth — leading to what I call a "ruminative cycle," one of the most common symptoms of depression. An early, well-publicized set of reports suggested that depressives are more realistic than happy people, but today a more considered view is that depressives are disordered in part because they place undue focus on negative things, often creating a downward spiral that is difficult to escape. Mark Twain once wrote, in a rare but perceptive moment of seriousness, "Nothing that grieves us can be called little; by the eternal laws of proportion a child's loss of a doll and a king's loss of a crown are events of the same size." Much, if not all, depression may begin with the magnification of loss, which in turn may stem directly from the ways in which memory is driven by context. Sad memories stoke sadder memories, and those generate more that are sadder still. To a person who is depressed, every fresh insult confirms a fundamental view that life is unfair or not worth living. Contextual memory thus stokes the memory of past injustices. (Meanwhile, motivated reasoning often leads depressives to discount evidence that would contradict their general view about the sadness of life.) Without some measure of self-control or a capacity to shift focus, the cycle may persist.

Such feedback cycles may even contribute a bit to bipolar disorder, not only in the "down" moments but also even in the manic ("up") phases. According to Kay Redfield Jamison, a top-notch psychologist who has herself battled manic depression, when one has bipolar disorder,

> there is a particular kind of pain, elation, loneliness, and terror involved in this kind of madness . . . When you're high it's tremendous. The ideas and feelings are fast and frequent like shoot-

ing stars . . . But, somewhere, this changes. The fast ideas are far too fast, and there are far too many; overwhelming confusion replaces clarity . . . madness carves its own reality.

Without sufficient inherent capacity for cognitive and emotional control, a bipolar person in a manic state may spiral upward so far that he or she loses touch with reality. Jamison writes that in one of her early manic episodes she found herself "in that glorious illusion of high summer days, gliding, flying, now and again lurching through cloud banks and ethers, past stars, and across fields of ice crystals . . . I remember singing 'Fly Me to the Moon' as I swept past those of Saturn, and thinking myself terribly funny. I saw and experienced that which had been only in dreams, or fitful fragments of aspiration." Manic moods beget manic thoughts, and the spiral intensifies.

Even the delusions common to schizophrenia may be exacerbated by — though probably not initially caused by — the effects of motivated reasoning and contextual memory. Many a schizophrenic, for example, has come to believe that he is Jesus and has then constructed a whole world around that notion, presumably "enabled" in part by the twin forces of confirmation bias and motivated reasoning. The psychiatrist Milton Rokeach once brought together three such patients, each of whom believed himself to be the Son of the Holy Father. Rokeach's initial hope was that the three would recognize the inconsistency in their beliefs and each in turn would be dissuaded from his own delusions. Instead, the three patients simply became agitated. Each worked harder than ever to preserve his own delusions; each developed a different set of rationalizations. In a species that combines contextually driven memory with confirmation bias and a strong need to construct coherent-seeming life narratives, losing touch with reality may well be an occupational hazard.

Depression (and perhaps bipolar disorder) is probably also aggravated by another one of evolution's glitches: the degree to which we depend on the somewhat quirky apparatus of pleasure. As we saw in the previous chapter, long before sophisticated deliberative reasoning arose, our pre-hominid ancestors presumably set their goals primarily by following the compass of pleasure (and avoiding its antithesis, pain). Even though modern humans have more sophisticated machinery for setting goals, pleasure and plain probably still form the core of our goal-setting apparatus. In depressives, this may yield a kind of double-whammy; in addition to the immediate pain of depression, another symptom that often arises is paralysis. Why? Quite possibly because the internal compass of pleasure becomes nonresponsive, leaving sufferers with little motivation, nothing to steer toward. For an organism that kept its mood separate from its goals, the dysfunction often accompanying depression might simply not occur.

In short, many aspects of mental illness may be traced to, or at least intensified by, some of the quirks of our evolution: contextual memory, the distorting effects of confirmation bias and motivated reasoning, and the peculiar split in our systems of self-control. A fourth contributor may be our species' thirst for explanation, which often leads us to build stories out of a sparse set of facts. Just as a gambler may seek to "explain" every roll of the dice, people afflicted with schizophrenia may use the cognitive machinery of explanation to piece together voices and delusions. This is not to say that people with disorders aren't different from healthy folks, but rather that their disorders may well have their beginnings in neural vulnerabilities that we all share.

Perhaps it is no accident, then, that so much of the advice given by cognitive-behavioral therapists for treating depression consists of getting people to cope with ordinary human failures in reasoning. David Burns's well-known *Feeling Good Handbook*, for example, suggests that ten basic cognitive errors, such as "overgeneralization" and "personalization," are made by people who are anxious or depressed. Overgeneralization is the process of erroneously "seeing a single event as a part of a never-ending pattern of defeat"; personalization is the mistake of assuming that we (rather than external events) are responsible for anything bad that happens. Both errors probably stem in part from the human tendency to extrapolate excessively from small amounts of highly salient data. One setback does not a miserable life make, yet it's human to treat the latest, worst news as an omen, as if a whole life of cyclical ups and downs is negated by a single vivid disaster. Such misapprehensions might simply not exist in a species capable of assigning equal mental weight to confirming and disconfirming evidence.

I don't mean to say that depression (or any disorder) is purely a byproduct of limitations in our abilities to objectively evaluate data, but the clumsy mechanics of our klugey mind very likely lay some of the shaky groundwork.

If disorders extend from fault lines, they certainly move beyond them too. To the extent that genes clearly play a role in mental disorders, evolution is in some way — adaptively or otherwise — implicated. But our mental fault lines sometimes give rise to earthquakes, though at other times only tiny tremblors, scarcely felt. Evolution, however haphazard, can't possibly be the whole story. Most common mental disorders seem to depend on a genetic component, shaped by evolution — but also on environmental causes that are not well understood. If one identical twin has, say, schizophrenia, the other one is considerably more likely than average to also have it, but the so-called "concordance" percentage, the chance that one twin will have the disorder if the other does, is only about 50 percent. For that reason alone it would clearly be overreaching to ascribe every aspect of mental illness to the idiosyncrasies of evolution.

But at the same time, it seems safe to say that no intelligent and compassionate designer would have built the human mind to be quite as vulnerable as it is. Our mental fragility provides yet another reason to doubt that we are the product of deliberate design rather than chance and evolution.

Which brings us to one last question, perhaps the most important of all: if the mind is a kluge, is there anything can we do about it?

8

True Wisdom

God grant me the serenity to accept the things I cannot change, courage to change the things I can, and wisdom to know the difference.

— Reinhold Niebuhr

To know that one knows what one knows, and to know that one doesn't know what one doesn't know, there lies true wisdom.

— Confucius

Human beings have intellectual skills of unparalleled power. We can talk, we can reason, we can dance, we can sing. We can debate politics and justice; we can work for the betterment not just of ourselves but our species. We can learn calculus and physics, we can invent, educate, and wax poetic. No other species comes close.

But not every advance has been to the good. The machinery of language and deliberative reason has led to enormous cultural and technological advances, but our brain, which developed over a billion years of pre-hominid ancestry, hasn't caught up. The bulk of our genetic material evolved before there was language, before there was explicit reasoning, and before creatures like us even existed. Plenty of rough spots remain.

In this book, we've discussed several bugs in our cognitive makeup: confirmation bias, mental contamination, anchoring, framing, inadequate self-control, the ruminative cycle, the focusing illusion, motivated reasoning, and false memory, not to mention absent-mindedness, an ambiguous linguistic system, and vulnerability to mental disorders. Our memory, contextually driven as it is, is ill suited to many of the demands of modern life, and our self-control systems are almost hopelessly split. Our ancestral mechanisms were shaped in a different world, and our more modern deliberative mechanisms can't shake the influence of that past. In every domain we have considered, from memory to belief, choice, language, and pleasure, we have seen that a mind built largely through the progressive overlay of technologies is far from perfect. None of these aspects of human psychology would be expected from an intelligent designer; instead, the only reasonable way to interpret them is as relics, leftovers of evolution.

In a sense, the argument I have presented here is part of a long tradition. Gould's notion of remnants of history, a key inspiration for this book, goes back to Darwin, who started his legendary work *The Descent of Man* with a list of a dozen "useless, or nearly useless" features — body hair, wisdom teeth, the vestigial tail bone known as the coccyx. Such quirks of nature were essential to Darwin's argument.

Yet imperfections of the mind have rarely been discussed in the context of evolution. Why should that be? My guess is that there are at least two reasons. The first, plain and simple, is that many of us just don't *want* human cognition to turn out to be less than perfect, either because it would be at odds with our beliefs (or fondest desires) or because it leads to a picture of humankind that we find unattractive. The latter factor arises with special force in scientific fields that try to characterize human behavior; the more we stubbornly deviate from rationality, the harder it is for mathematicians and economists to capture our choices in neat sets of equations.

A second factor may stem from the almost mystifying popularity of creationism, and its recent variant, intelligent design. Few theories are as well supported by evidence as the theory of evolution, yet a large portion of the general public refuses to accept it. To any scientist familiar with the facts — ranging from those garnered through the painstaking day-to-day studies of evolution in the contemporary Galápagos Islands (described in Jonathan Weiner's wonderful book *The Beak of the Finch*) to the details of molecular change emerging from the raft of recently completed genomes — this continued resistance to evolution seems absurd.* Since so much of it seems to come from people who have trouble accepting the notion that well-organized structure could have emerged without forethought, scientists often feel compelled to emphasize evolution's high points — the cases of well-organized structure that emerged through sheer chance.

Such emphasis has led to a great understanding of how a blind process like evolution can produce systems of tremendous beauty — but at the expense of an equally impassioned exploration of the illuminating power of imperfection. While there is nothing inherently wrong in examining nature's greatest hits, one can't possibly get a complete and balanced picture by looking only at the highlights.

The value of imperfections extends far beyond simple balance, however. Scientifically, every kluge contains a clue to our past; wherever there is a cumbersome solution, there is insight into how nature layered our brain together; it is no exaggeration to say that the history of evolution is a history of overlaid technologies, and kluges help expose the seams.

Every kluge also underscores what is fundamentally wrongheaded about creationism: the presumption that we are the product of an all-seeing entity. Creationists may hold on to the bitter end, but imperfection (unlike perfection) beggars the imagination. It's one thing to imagine an all-knowing engineer designing a perfect eyeball, another to imagine that engineer slacking off and building a half-baked spine.

There's a practical side too: investigations into human idiosyn-

---

*It is sometimes said, pejoratively, that evolution is "just a theory," but this statement is true only in the technical sense of the word *theory* (that is, evolution is an explanation of data), not in the lay sense of being an idea about which there is reasonable doubt.

crasy can provide a great deal of useful insight into the human condition; as they say in Alcoholics Anonymous, recognition is the first step. The more we can understand our clumsy nature, the more we can do something about it.

When we look at imperfections as a source of insight, the first thing to realize is that not every imperfection is worth fixing. I've long since come to terms with the fact that my calculator is better than I am at solving square roots, and I see little point in cheering for Garry Kasparov over his computer opponent, Deep Blue, in the world chess championships. If computers can't beat us now at chess and Trivial Pursuit, they will someday soon. John Henry's fin-de-siècle Race Against the Machine was noble but, in hindsight, a lost cause. In many ways machines have (or eventually will have) the edge, and we might as well accept it. The German chemist Ernst Fischer mused that "as machines become more and more efficient and perfect, so it will become clear that imperfection is the greatness of man." A creature designed by an engineer might never know love, never enjoy art, never see the point of poetry. From the perspective of brute rationality, time spent making and appreciating art is time that could be "better" spent gathering nuts for winter. From my perspective, the arts are part of the joy of human existence. By all means, let us make poetry out of ambiguity, song and literature out of emotion and irrationality.

That said, not every quirk of human cognition ought to be celebrated. Poetry is good, but stereotyping, egocentrism, and our species-wide vulnerability to paranoia and depression are not. To accept *everything* that is inherent to our biological makeup would be to commit a version of the "naturalistic fallacy," confusing what is natural with what is good. The trick, obviously, is to sort through our cognitive idiosyncrasies and decide which are worth addressing and which are worth letting go (or even celebrating).

For example, it makes little sense to worry about ambiguity in everyday conversation because we can almost always use context and interaction to figure out what our conversational partners have in mind. It makes little sense to try to memorize the phone numbers of everyone we know because our memory just isn't built that way (and now we have cell phones to do that for us). For much of our daily business, our mind is more than sufficient. It generally keeps us well-fed, employed, away from obstacles, and out of harm's reach. As much as I envy the worry-free life of the average domesticated cat, I wouldn't trade my brain for Fluffy's for all the catnip in China.

But that doesn't mean that we can't, as thinkers, do even better. In that spirit, I offer, herewith, 13 suggestions, each founded on careful empirical research:

1. Whenever possible, consider alternative hypotheses. As we have seen, we humans are not in the habit of evaluating evidence in dispassionate and objective ways. One of the simplest things we can do to improve our capacity to think and reason is to discipline ourselves to consider alternative hypotheses. Something as simple as merely forcing ourselves to list alternatives can improve the reliability of reasoning.

One series of studies has shown the value of the simple maxim "Consider the opposite"; another set has shown the value of "counterfactual thinking" — contemplating what might have been or what could be, rather than focusing on what currently is.

The more we can reflect on ideas and possibilities other than those to which we are most attached, the better. As Robert Rubin (Bill Clinton's first treasury secretary) said, "Some people I've encountered in various phases of my career seem more certain about everything than I am about anything." Making the right choice often requires an understanding of the road not traveled as well as the road ultimately taken.

2. Reframe the question. Is that soap 99.4 percent pure or 0.6 percent toxic? Politicians, advertisers, and even our local supermarket staff routinely spin just about everything we hear, see, and read. Everything is presented to be as positive as possible. Our job — as consumers, voters, and citizens — must be to perpetually cast a skeptical eye and develop a habit of rethinking whatever we are asked. (Should I construe this "assisted suicide" legislation as an effort to protect people from murderous doctors or as a way of allowing folks to die with dignity? Should I think about the possibility of reducing my hours to part-time work as a pay cut or as an opportunity to spend more time with my kids?) If there's another way to think about a problem, do it. Contextual memory means that we are always swimming upstream: how we think about a question invariably shapes what we remember, and what we remember affects the answers we reach. Asking every question in more than one way is a powerful way to counter that bias.

3. Always remember that correlation does not entail causation. Believe it or not, if we look across the population of the United States, shoe size is highly correlated with general knowledge; people with bigger shoes tend to know more history and more geography than people with smaller shoes. But that doesn't mean buying bigger shoes will make you smarter, or even that having big feet makes you smart. This correlation, like so many others, seems more important than it really is because we have a natural tendency to confuse correlation with causation. The correlation I described is real, but the natural inference — that one factor must be causing the other — doesn't follow. In this example, the reason that the correlation holds is that the people with the littlest feet (and tiniest shoes) are our planet's newest visitors: infants and toddlers, human beings too young to have yet taken their first history class. We learn as we grow, but that doesn't mean that growing (per se) makes us learn.*

---

*Pop quiz: should we study the dictionary to make ourselves smarter? Maybe, maybe not: lots of websites promising to build vocabulary tell us that "people with bigger vocabularies are more successful," but is it the vocabulary that makes them successful, or some third factor, like intelligence or dedication, that leads to both success and a large vocabulary?

4. Never forget the size of your sample. From medicine to baseball statistics, people often fail to take into account the amount of data they've used in drawing their conclusions. Any single event may be random, but recurrence of the same pattern over and again is less likely to be an accident. Mathematically speaking, the bigger the sample, the more reliable the estimate. That's why, on average, a poll of 2,000 people is a lot more reliable than a poll of 200 people, and seeing someone bat .400 (successfully getting a hit in 40 percent of their tries) over 10 baseball games doesn't mean nearly as much as seeing them bat .400 over a 162-game season.

As obvious as this fact is, it's easy to forget. The person who first formalized this notion, known as the law of large numbers, thought it was so obvious that "even the stupidest man knows [it] by some instinct of nature," yet people routinely ignore it. We can't help but search for "explanations" of patterns in our data, even in small samples (say, a handful of baseball games or a single day's stock market results) that may well reflect nothing more than random chance. Boomer hit .400 in the last ten games because "he's seeing the ball real well," never because (statistically speaking) a .300 hitter is likely to occasionally look like a .400 hitter for a few days. Stock market analysts do the same thing, tying every day's market moves to some particular fact of the news. "The market went up today because Acme Federated reported unexpectedly high fourth-quarter results." When was the last time you heard any analyst say "Actually, today's rise in the market was probably nothing more than a random fluctuation"?

Happily, psychologist Richard Nisbett has shown that ordinary folks can be taught to be more sensitive to the law of large numbers in less than half an hour.

5. Anticipate your own impulsivity and pre-commit. Odysseus tied himself to a mast to resist the temptations of the Sirens; we would all do well to learn from him. Compare, for example, the groceries we might choose a week in advance, with a well-rested stomach, to the junk we buy in the store when we are hungry. If we commit ourselves in advance to purchasing only what we've decided on in advance, we come home with a more healthful basket of groceries. "Christmas Clubs," which tie up money all year long for holiday shopping, are completely irrational from the perspective of an economist — why earmark money when liquidity is power? — but become completely sensible once we acknowledge our evolved limitations. Temptation is greatest when we can see it, so we are often better off in plans for the future than in impulses of the moment. The wise person acts accordingly.

6. Don't just set goals. Make contingency plans. It's often almost impossible for people to stick to vague goals like "I intend to lose weight" or "I plan to finish *this* article before the deadline." And simply making a goal more specific ("I plan to lose *six* pounds") is not enough. But research by the psychologist Peter Gollwitzer shows that by transforming goals into specific *contingency plans* — of the form "if X, then Y" (for example, "If I see French fries, then I will avoid them") — we can markedly increase the chance of success.

A recognition of our klugey nature can help explain why our late-evolved deliberative reasoning, grafted onto a reflexive, ancestral system, has limited access to the brain's steering wheel; instead, almost everything has to pass through the older ancestral, reflexive system. Specific contingency plans offer a way of working around that limitation by converting abstract goals into a format (*if-then*, basic to all reflexes) that our ancestral systems can understand. To the extent that we can speak the language of our older system, we increase our chances of achieving our goals.

7. Whenever possible, don't make important decisions when you are tired or have other things on your mind. Thinking while tired (or distracted) is not so different from driving while drinking. As we get tired, we rely more on our reflexive system, less on deliberative reasoning; ditto as we get distracted. One study, for example, showed that a healthful-minded consumer who is given a choice between a fruit salad and a chocolate cake becomes more likely to choose the cake when forced to remember a seven-digit number. If we want to reason by emotion alone, fine, but if we prefer rationality, it is important to create "winning conditions" — and that means, for important decisions, adequate rest and full concentration.

8. Always weigh benefits against costs. Sounds obvious, but it is not something that comes naturally to the human mind. People tend to find themselves in either a "prevention" frame of mind, emphasizing the costs of their actions (if I don't go, I'll waste the money I spent on concert tickets), or a "promotion" frame of mind, emphasizing the benefits (It'll be fun! Who cares if I'll be late for work in the morning?). Sound judgment obviously requires weighing both costs and benefits, but unless we are vigilant, our temperament and mood often stand in the way.

Pay special attention, by the way, to what some economists call "opportunity costs"; whenever you make an investment, financial or otherwise, ponder what else you might be doing instead. If you're doing one thing, you can't do another — a fact that we often forget. Say, for example, that people are trying to decide whether it makes sense to invest $100 million in public funds in a baseball stadium. That $100 million may well bring some benefits, but few people evaluate such projects in the context of what *else* that money might do, what opportunities (such as paying down the debt to reduce future interest payments or building three new elementary schools) must be foresworn in order to make that stadium happen. Because such costs don't come with a readily visible price tag, we often ignore them. On a personal level, taking opportunity costs into account means realizing that whenever we make a choice to do something, such as watch television, we are using time that could be spent in other ways, like cooking a nice meal or taking a bike ride with our kids.

9. Imagine that your decisions may be spot-checked. Research has shown that people who believe that they will have to justify their answers are less biased than people who don't. When we expect to be held accountable for our decisions, we tend to invest more cognitive effort and make correspondingly more sophisticated decisions, analyzing information in more detail.

For that matter (and no, I'm not making this up) office workers are more likely to pay for coffee from a communal coffee machine if the coffee machine is positioned under a poster featuring a pair of eyes — which somehow makes people feel that they are accountable — than under a poster that has a picture of flowers.

10. Distance yourself. Buddhists tells us that everything seems more important in the moment, and for the most part, they're right. If an out-of-control car is bearing down on you, by all means, drop everything and focus all of your energies on the short-term goal of getting out of the way. But if I want to top off the meal with that chocolate cake, I should ask myself this: am I overvaluing my current goals (satisfying my sweet tooth) relative to my long-term goals (staying healthy)? It'll feel good now to send that email excoriating your boss, but next week you'll probably regret it.

Our mind is set up to ponder the near and the far in almost totally different ways, the near in concrete terms, the far in abstract terms. It's not *always* better to think in more distant terms; remember the last time you promised to do something six months hence, say, attend a charity event or volunteer at your child's school? Your promise probably seemed innocuous at the time but might have felt like an imposition when the date came to actually fulfill it. Whenever we can, we should ask, How will my future self feel about this decision? It pays to recognize the differences in how we treat the here and now versus the future, and try to use and balance both modes of thinking — immediate and distant — so we won't fall prey to basing choices entirely on what happens to be in our mind in the immediate moment. (A fine corollary: wait awhile. If you still want it tomorrow, it may be important; if the need passes, it probably wasn't.) Empirical research shows that irrationality often dissipates with time, and complex decisions work best if given time to steep.

11. Beware the vivid, the personal, and the anecdotal. This is another corollary to "distancing ourselves," also easier said than done. In earlier chapters we saw the relative temptation prompted by cookies that we can see versus cookies that we merely read about. An even more potent illustration might be Timothy Wilson's study of undergraduates and condom brands, which yielded a classic "do as I say, not as I do" result. Subjects in the experiment were given two sources of information, the results of a statistically robust study in *Consumer Reports* favoring condoms of Brand A and a single anecdotal tale (allegedly written by another student) recommending Brand B, on the grounds that a condom of Brand A had burst in the middle of intercourse, leading to considerable anxiety about possible pregnancy. Virtually all students agreed *in principle* that *Consumer Reports* would be more reliable and also that they would not want their friends to choose on the basis of anecdotal evidence. But when asked to choose for themselves, nearly a third (31 percent) still yielded to the vivid and anecdotal, and went with Brand B. Our four-legged ancestors perhaps couldn't help but pay attention to whatever seemed most colorful or dramatic; we have the luxury to take the time to reflect, and it behooves us to use it, compensating for our vulnerability to the vivid by giving special weight to the impersonal but scientific.

12. Pick your spots. Decisions are psychologically, and even physically, costly, and it would be impossible to delay every decision until we had complete information and time to reflect on every contingency and counteralternative. The strategies I've given in this list are handy, but never forget the tale of Buridan's Ass, the donkey that starved to death while trying to choose between two equally attractive, equally close patches of hay. Reserve your most careful decision making for the choices that matter most.

13. Try to be rational. This last suggestion may sound unbelievably trivial, on par with the world's most worthless stock market advice ("Buy low, sell high" — theoretically sound yet utterly useless). But reminding yourself to be rational is not as pointless as it sounds.

Recall, for example, "mortality salience," a phenomenon I described earlier in the chapter on belief: people who are led in passing to think about their own death tend to be harsher toward members of other groups. Simply telling them to consider their answers before responding and "to be as rational and analytic as possible" (instead of just answering with their "gut-level reactions") reduces the effect. Another recent study shows similar results.

One of the most important reasons why it just might help to tell yourself to be rational is that in so doing, you can, with practice, automatically *prime* yourself to use some of the other techniques I've just described (such as considering alternatives or holding yourself accountable for your decisions). Telling ourselves to be rational isn't, on its own, likely to be enough, but it might just help in tandem with the rest.

Every one of these suggestions is based on sound empirical studies of the limits of the human mind. Each, in its own way, addresses a different weakness in the human mind and each, in its own way, offers a technique for smoothing out some of the rough spots in our evolution.

With a properly nuanced understanding of the balance between the strengths and weaknesses of the human mind, we may have an opportunity to help not only ourselves but society. Consider, for example, our outmoded system of education, still primarily steeped in ideas from nineteenth-century pedagogy, with its outsized emphasis on memorization echoing the Industrial Revolution and Dickens's stern schoolmaster, Mr. Gradgrind: "Now, what I want is, Facts. Teach these boys and girls nothing but Facts . . . Plant nothing else, and root out everything else." But it scarcely does what education ought to do, which is to help our children learn how to fend for themselves. I doubt that such a heavy dose of memorization *ever* served a useful purpose, but in the age of Google, asking a child to memorize the state capital has long since outlived its usefulness.

Deanna Kuhn, a leading educational psychologist and author of the recent book *Education for Thinking,* presents a vignette that reminds me entirely too much of my own middle-school experience: a seventh-grader at a considerably above average school asked his (well-regarded) social studies teacher, "Why do we have to learn the names of all thirteen colonies?" The teacher's answer, delivered without hesitation, was "Well, we're going to learn all fifty states by June, so we might as well learn the first thirteen now." Clear evidence that the memorization cart has come before the educational horse. There is value, to be sure, in teaching children the history of their own country and — especially in light of increasing globalization — the world, but a memorized list of states casts no real light on history and leaves a student with no genuine skills for understanding (say) current events. The result, in the words of one researcher, is that

> many students are unable to give evidence of more than a superficial understanding of the concepts and relationships that are fundamental to the subjects they have studied, or of an ability to apply the content knowledge they have acquired to real-world problems . . . it is possible to finish 12 or 13 years of public education in the United States without developing much competence as a thinker.

In the information age, children have no trouble *finding* information, but they have trouble *interpreting* it. The fact (discussed earlier) that we tend to believe first and ask questions later is truly dangerous in the era of the Internet — wherein anyone, even people with no credentials, can publish anything. Yet studies show that teenagers frequently take whatever they read on the Internet at face value. Most students only rarely or occasionally check to see who the author of a web page is, what the person's credentials are, or whether other sources validate the information in question. In the words of two Wellesley college researchers, "Students use the Net as a primary source of information, usually with little regard as to the accuracy of that information." The same is true for most adults; one Internet survey reported that "the average consumer paid far more attention to the superficial aspects of a [web] site, such as visual cues, than to its content. For example, nearly half of all consumers (or 46.1%) in the study assessed the credibility of sites based in part on the appeal of the overall visual design of a site, including layout, typography, font size, and color schemes."*

Which is exactly why we need schools and not just Wikipedia and an Internet connection. If we were naturally good thinkers, innately skeptical and balanced, schools would be superfluous.

But the truth is that without special training, our species is inherently gullible. Children are born into a world of "revealed truths," where they tend to accept what they are told as gospel truth. It takes work to get children to understand that often multiple opin-

---

*Another study, conducted before the Web existed as such, pointed in the same direction. Educational psychologist David Perkins asked people with a high school or college education to evaluate social and political questions, such as "Does violence on television significantly increase the likelihood of violence in real life?" or "Would restoring the military draft significantly increase America's ability to influence world events?"; responses were evaluated in terms of their sophistication. How many times did people consider objections to their own arguments? How many different lines of argument did people consider? How well could people justify their main arguments? Most subjects settled for simplistic answers no matter how much the experimenters tried to push them — and the amount of education people had made surprisingly little difference. As Perkins put it, "Present educational practices do little to foster the development of informal reasoning skills."

ions exist and that not everything they hear is true; it requires even more effort to get them to learn to evaluate conflicting evidence. Scientific reasoning is not something most people pick up naturally or automatically.

And, for that matter, we are not born knowing much about the inner operations of our brain and mind, least of all about our cognitive vulnerabilities. Scientists didn't even determine with certainty that the brain was the source of thinking until the seventeenth century. (Aristotle, for one, thought the purpose of the brain was to cool the blood, inferring this backward from the fact that large-brain humans were less "hot-blooded" than other creatures.) Without lessons, we are in no better position to understand how our mind works than how our digestive system works. Most us were never taught how to take notes, how to evaluate evidence, or what human beings are (and are not) naturally good at. Some people figure these things out on their own; some never do. I cannot recall a single high school class on informal argument, how to spot fallacies, or how to interpret statistics; it wasn't until college that anybody explained to me the relation between causation and correlation.

But that doesn't mean we *couldn't* teach such things. Studies in teaching so-called critical thinking skills are showing increasingly promising results, with lasting effects that can make a difference. Among the most impressive is a recent study founded on a curriculum known as "Philosophy for Children," which, as its name suggests, revolves around getting children to think about — and discuss — philosophy. Not Plato and Aristotle, mind you, but stories written for children that are explicitly aimed at engaging children in philosophical issues. The central book in the curriculum, *Harry Stottlemeier's Discovery* (no relation to Harry Potter), begins with a section in which the eponymous Harry is asked to write an essay called "The Most Interesting Thing in the World." Harry, a boy after my own heart, chooses to write his on thinking: "To me, the most interesting thing in the whole world is thinking. I know that lots of other things are also very important and wonderful, like electricity, and magnetism and gravitation. But although we understand them, they can't understand us. So thinking must be something very special."

Kids of ages 10–12 who were exposed to a version of this curriculum for 16 months, for just an hour a week, showed significant gains in verbal intelligence, nonverbal intelligence, self-confidence, and independence.

Harry Stottlemeier's essay — and the "Philosophy for Children" curriculum — is really an example of what psychologists call *metacognition,* or knowing about knowing. By asking children to reflect on how they know what they know, we may significantly enhance their understanding of the world. Even a single course — call it "The Human Mind: A User's Guide" — could go a long way.

No such guide will give us the memory power to solve square roots in our head, but many of our cognitive peccadilloes *are* addressable: we can train ourselves to consider evidence in a more balanced way, to be sensitive to biases in our reasoning, and to make plans and choices in ways that better suit our own long-term goals. If we do — if we learn to recognize our limitations and address them head on — we just might outwit our inner kluge.

Acknowledgments

Amanda Cook, editor of all editors, is a genius with vision who often left me feeling the sort of joy that actors must get when working for a great director. Amanda helped conceive the book and shepherded me through three exacting revisions. As if that weren't enough, I also got fantabulous editorial advice from Neil Belton, my British editor; Don Lamm, half of the team that helped set me up with Amanda and Neil in the first place; and my wife, Athena, who, when it comes to editing, is an amateur with the skills of a professional. It's hard to imagine another author being so lavished in editorial wisdom.

Conceptual wisdom came from a host of friends and colleagues. Zach Woods, Yaacov Trope, Hugh Rabagliati, Athena Vouloumanos, Rachel Howard, Iris Berent, Ezequiel Morsella, Cedric Boeckx, Deanna Kuhn, Erica Roedder, Ian Tattersall, and two sets of students at NYU generously read and critiqued the complete manuscript, while Meehan Crist, Andrew Gerngross, Joshua Greene, George Hadjipavlou, Jon Jost, Steve Pinker, and my father, Phil Marcus, made penetrating comments on individual chapters. I also thank Scott Atran, Noam Chomsky, Randy Gallistel, Paul Glimcher, Larry Maloney, and Massimo Piatelli-Palmarini for helpful discussion. Numerous people, some whom I've never met, helped me with queries ranging from the syntax of Esperanto to the evolution of the eyes of animals and the carbon cycle of plants; these include Don Harlow, Lawrence Getzler, Tyler Volk, Todd Gureckis, Mike Landy, and Dan Nilsson; my apologies to those I've failed to thank. I have only my memory to blame.

Christy Fletcher and Don Lamm are the dynamic duo who helped sell this book and connect me with Amanda Cook and Neil Belton; they've been supportive, energetic, and involved, everything an agent (or pair of agents) should be.

Finally, I'd like to thank my family — especially Athena, Mom, Dad, Linda, Julie, Peg, Esther, Ted, and Ben, and my in-laws Nick, Vickie, and the Georges — for their enthusiasm and unstinting support. Writing can be hard work, but with so many talented and loving people behind me, it's always a pleasure.

Notes

1. Remnants of History

2   The average person can't keep a list of words straight for a half an hour: Tulving & Craik, 2000.
5   One scientist: Wesson, 1991.
7   "Human cognition approaches an optimal level of performance": Chater et al., 2006.
    Superlatively well engineered functional designs: Tooby & Cosmides, 1995.
    In principle possibility of "inept evolution": Tooby & Cosmides, 1995.
8   *The Selfish Gene:* Dawkins, 1976.
    Infanticide: Daly & Wilson, 1988.
    Male overperception of female sexual intent: Haselton & Buss, 2000.
9   Evolution as mountain climbing: Dawkins, 1996.
10 Bar-headed goose: Fedde et al., 1989.
    No guarantee that evolution will ever reach the *highest* peak: Dawkins, 1982.
11 The inefficiency of the gaps across which neurons communicate: Montague, 2006.
12 New genes in concert with old genes: Marcus, 2004.
    Evolution like a tinkerer: Jacob, 1975.
13 Hindbrain evolution: Rosa-Molinar et al., 2005.
    Midbrain evolution: Takahashi, 2005.
    Language and the brain: Gebhart et al., 2002; Demonet et al., 2005.
14 "Progressive overlay of technologies": Allman, 1999.

Chimpanzee overlap: The Chimpanzee Sequencing and Analysis Consortium 2005; King & Wilson, 1975.

15 *How Doctors Think:* Groopman, 2007.
    The March of Folly: Tuchman, 1984.

16 Underestimating Mother Nature: Dennett, 1995.

2. Memory

18 Teenagers and World War I: Kelly, 2001.

19 Time spent looking for lost items: Tyre, 2004.
    Skydivers who forgot to pull the ripcord: http://temagami.carleton.ca/jmc/cnews/18112005/n6.shtml.

22 Remembering the frequent, the recent, and the relevant: Anderson, 1990.

24 Study underwater, test underwater: Godden & Baddeley, 1975.

25 Walking more slowly to the elevator: Bargh et al., 1996.
    Soccer hooligans: Dijksterhuis & van Knippenberg, 1998.
    Minority groups, priming, and testing: Steele & Aronson, 1995.
    The automatic nature of stereotype priming: Greenwald et al., 1998.

26 Random dot patterns: Posner & Keefe, 1968.

27 9/11 memories: Talarico & Rubin, 2003.

28 Faulty eyewitness testimony: Schacter, 2001.

30 Memory and lability: Debiec et al., 2006.

33 Reconstructive memory for events: Loftus, 2003.
    Mere exposure and confusions about fame: Jacoby et al., 1989.

34 How an actor's memory works: Noice & Noice, 2006.

35 Photographic memory: http://www.slate.com/id/2140685/.

36 Pinker, quoted on book jacket of Schacter, 2001.

37 Fractured memory and preparing for the future: Schacter & Addis, 2007.

3. Belief

40 Gullibility: Forer, 1949.

42 Snowball study: Dion, 1972.
    Beauty studies: Etcoff, 1999.
    Candidates that look more competent: Todorov et al., 2005.

43 Would you like carrots with that? (food in McDonald's wrap): Robinson et al., 2007.

44 Impressions of Donald: Higgins et al., 1977.
    The focusing illusion and dating: Strack, Martin, & Schwarz, 1988.
46 Dishes and perceived contribution: Leary & Forsyth, 1987; Ross & Sicoly, 1979.
    Wheel of fortune: Tversky & Kahneman, 1974.
    Attila the Hun: Russo & Schoemaker, 1989.
47 Limit 12 per customer: Wansink et al., 1998.
48 Lips: Strack, Martin, & Stepper, 1988.
    Arm flexion: Förster & Strack, 1998.
    Chinese characters: Zajonc, 1968.
    Name letter effect: Nuttin, 1987.
    Mere exposure and paintings: James Cutting, personal communication, based on informal classroom research.
50 Thinking about death: Solomon et al., 2004.
    Minority groups during crisis: Jost & Hunyadi, 2003.
53 The difficulty in repressing automatic thoughts: Wegner, 1994; Macrae et al., 1994.
    Sequences: Wason, 1960.
54 Another study: Darley & Gross, 1983.
55 Trivia game: Dijksterhuis & van Knippenberg, 1998.
56 Love the one you'll be with: Berscheid et al., 1976.
    Motivated reasoning: Kunda, 1990.
57 Cigarettes and rationalization: Kassarjian & Cohen, 1965.
    Critiques of studies that challenge our prior beliefs: Lord et al., 1979.
58 Bush: "I trust God speaks through me. Without that, I couldn't do my job." July 9, 2004, Lancaster, PA: http://www.freerepublic.com/focus/f-news/1172948/posts.
    Voting preferences and belief in God: Pew Research Center, 2007.
    Belief in a just world: Lerner, 1980.
61 Trouble telling sound arguments from fallacies: Stanovich, 2003.
62 Confusing logic with prior beliefs: Klauer et al., 2000; Oakhill et al., 1989.
63 Neural basis of syllogism: Goel, 2003; Goel & Dolan, 2003.
    Understanding of syllogisms in other cultures: Luria, 1971.
64 Author's study on belief: Marcus, 1989.
66 Interruption and gullibility: Gilbert et al., 1990.
    Time pressure and cognitive strain increase the chance of believing falsehoods: Gilbert et al., 1993.
    This American Life: April 13, 2007: http://www.thislife.org/Radio_Episode.aspx?sched=1183).

Two-month investigation: http://www.nhpr.org/node/12381.
67 Ask and ye shall believe: Pandelaere & Dewitte, 2006.

4. CHOICE

69 Kids and marshmallows test: Mischel et al., 1989.
70 $100 now or $300 in three years: Ainslie, 2001.
    Day 89 of 90: Butler et al., 1995.
73 Almost everyone takes the sure thing: Allais, 1953.
74 Drive across town: Tversky & Kahneman, 1981.
77 Retirement: http://archives2.sifma.org/research/pdf/RsrchRprtVol7-7.pdf.
79 Satisfaction and cost: Thaler, 1999.
80 Shopkeeper's tale: Cialdini, 1993.
    Anchoring in insurance for the intangible: Jones-Lee & Loomes, 2001.
81 Framing: Tversky & Kahneman, 1981.
    96.3 percent crime-free: Quattrone & Tversky, 1988.
82 Prevention and promotion: Higgins, 2000.
83 The choices made by hungry people: Read & van Leeuwen, 1998.
85 Credit card debt: Aizcorbe et al., 2003.
    Future discounting and the uncertainty of the ancestral world: Kagel et al., 1986.
88 Farm workers versus dolphins: Kahneman & Ritov, 1994.
    Lemon-lime study: Winkielman & Berridge, 2004.
    Prisoner's dilemma, contaminated by news broadcasts: Hornstein et al., 1975.
89 Seeing and smelling the cookies: Ditto et al., 2006.
    Caution to the wind: Ditto et al., 2006.
    Attractiveness and the perception of risk: Blanton & Gerrard, 1997.
90 Trolley problem: Greene et al., 2001; Thomson, 1985.
    Christmas truce: Brown & Seaton, 1984.
91 Moral intuitions: Haidt, 2001.
92 fMRI studies of moral dilemmas: Greene et al., 2004; Greene et al., 2001.

5. LANGUAGE

96 Plane crash: Cushing, 1994.
97 Russell quotation: Russell, 1918.

104 Loglan, short for "logical language": Brown & Loglan Institute, 1975.

105 Evolution of speech: Lieberman, 1984.

108 Tongue-twisters and timing mechanisms: Goldstein et al., 2007.

112 Generics: Gelman & Bloom, 2007; Prasada, 2000.

113 Generics and our split reasoning systems: Leslie, 2007.

114 "What some super-engineer would construct": Chomsky, 2000.
     Chomsky's effort to capture language with a small set of laws: Chomsky, 1995.

115 Book on physics: Smolin, 2006.

116 Chomsky and colleagues: Hauser et al., 2002.
     Pinker and Jackendoff: Pinker & Jackendoff, 2005.
     Chimpanzee language is missing more than just recursion: Premack, 2004.
     The trouble with trees: Marcus & Wagers, under review.

117 Troubles with center embedding: Miller & Chomsky, 1963.
     Types of recursion: Parker, 2006.

118 Problem with trees: Parker, 2006.

120 Study of ambiguity: Keysar, 2002.

121 "Good enough" language: Ferreira et al., 2002.
     Animals on the ark: Reder & Kusbit, 1991.
     More people have been to Russia than I have: Montalbetti, 1984.

6. PLEASURE

124 237 reasons to have sex: Meston & Buss, 2007.
     Automatic classification into good and bad: Fazio, 1986.

125 The more we need 'em, the more we like 'em: Fishbach et al., 2004.
     Leisure time: U.S. Department of Labor Statistics, 2007.

126 Television watching: Kahneman et al., 2004; Nielsen, 2006.

127 Reaching for the whiskey bottle: Cheever, 1990.
     Primrose path: Herrnstein & Prelac, 1992.

132 Video game design: Thompson, 2007.

133 Lullaby hypothesis: Trehub, 2003.
     Music and sexual selection: Miller, 2000.
     Problems with theories of music evolution: Fitch, 2005.

134 Apes that beat their chests rhythmically: Fitch, 2005.
     Goldfish and pigeons trained to distinguish musical styles: Fitch, 2005.

137 Coping ability in victims of accidents: Brickman et al., 1978; Linley & Joseph, 2004.

138 The relation between wealth and happiness: Kahneman et al., 2006.
Happiness in Japan, 1958 to 1987: Easterlin, 1995.
Happiness in the United States: Easterlin, 1995.
Happiness and money: Layard, 2005.
Relative income: Frank, 2001.
The happiness (hedonic) treadmill: Brickman & Campbell, 1971.

139 Dates: Strack et al., 1988.
Marriages: Schwarz, 1991.
Health: Smith et al., 2006.
The more we think about how happy we are: Ariely et al., 2006.
Rumination: Lyubomirsky et al., 1998.

141 Smarter, fairer, more considerate, more dependable, more creative: Alicke et al., 1995; Brown, 1986; Dunning et al., 1989; Messick et al., 1985.
Better drivers: Svenson, 1981.
Better than average health: Kirscht et al., 1966.
Cognitive dissonance: Festinger & Carlsmith, 1959.

142 Serotonin evolution: Allman, 1999.
Emotions across species: Ledoux, 1996.

143 Souped-up anterior cingulate: Allman et al., 2002.
Anterior cingulate and conflict between cognitive systems: McClure et al., in press.
Teenagers: Galvan et al., 2006.
Evolutionary old before new: Finlay & Darlington, 1995.

7. THINGS FALL APART

145 Eskimo words for snow: Pullum, 1991.
Neural noisiness: Montague, 2006.
Junk food and cognitive load: Shiv & Fedorikhin, 1999.

146 Deliberative systems left behind: Ferreira et al., 2006.
Stereotyping and cognitive load: Sherman, 2000.
Egocentrism and cognitive load: Epley et al., 2004.
Anchoring and cognitive load: Epley & Gilovich, 2006.

147 Daydreaming about sex: Howard, 2004.
Zoning out: Schooler et al., 2004.

Recent NHTSA study: http://www.nhtsa.dot.gov/staticfiles/DOT/NHTSA/ NRD/Multimedia/PDFs/Crash%20Avoidance/2006/DriverInattention .pdf.

Leading causes of death: Minino et al., 2002.

148 Costs of procrastination: Steel, 2007.

The "quintessential self-regulatory failure": Steel, 2007.

149 Lifetime prevalence of disorders: Kessler et al., 2005.

151 Vonnegut, tribes, and extended families: National Public Radio interview, January 23, 2006, http://www.npr.org/templates/story/story .php?storyId=5165342.

Virtually every trait depends on a mix of factors, some environmental, some biological: Marcus, 2004; Plomin, 1997.

Shamans and schizophrenia: Polimeni & Reiss, 2002.

Agoraphobia: Nesse, 1997.

Anxiety: Marks & Nesse, 1997.

Depression: Price et al., 1997

Schizophrenia incidence: Goldner et al., 2002.

152 Malaria: Nesse & Williams, 1994.

Sociopathy: Mealey, 1995.

155 Cost-benefit and orbitofrontal damage: Bechara et al., 2000.

157 Depressive realism: Alloy & Abramson, 1979; Pacini, 1998.

Mood-congruent memory: Watkins et al., 1996.

160 Most mental disorders have a genetic component: Plomin et al., 2001.

8. True Wisdom

162 Imperfections of the mind rarely discussed: For three notable exceptions, see Clark, 1987; Linden, 2007; and Stitch (in press).

163 *The Beak of the Finch:* Weiner, 1994.

Recently completed genomes: http://www.ensembl.org/index.html; Gregory, 2005.

165 Listing alternatives: Hoch, 1985; Koriat et al., 1980.

Consider the opposite: Arkes, 1991; Larrick, 2004; Mussweiler et al., 2000.

Counterfactual thinking: Galinsky & Moskowitz, 2000; Kray et al., 2006.

167 Sample size education: Nisbett et al., 1983.

168 Precommitment: Lynch & Zauberman, 2006.
Specific contingency plans: Gollwitzer & Sheeran, 2006.
169 Fruit salad versus cake, under cognitive load: Shiv & Fedorikhin, 1999.
170 Accountability: Tetlock, 1985.
Eyes and the honor system: Bateson et al., 2006.
Distance and decision making: Liberman et al., 2002; Lynch & Zauberman, 2006.
171 Irrationality dissipating with time: Koehler, 1994.
Time to steep: Dijksterhuis & Nordgren, 2006.
Anecdotal data, against better judgment: Wilson & Brekke, 1994.
Psychological cost of decisions: Schwartz & Schwartz, 2004.
Physical cost of decisions: Gailliot et al., 2007.
172 Mortality salience and the effects of being told to "be rational": Simon et al., 1997. See also Epstein et al., 1992.
Further evidence that admonitions to be rational might work: Ferreira et al., 2006.
173 *Education for Thinking:* Kuhn, 2005.
Graduating without much competence as a thinker: Nickerson, 1988.
174 Taking the Internet at face value: Metzger et al., 2003.
With little regard as to the accuracy: Graham & Metaxas, 2003.
Credibility and visual design: Fogg et al., 2002.
Unsophisticated evaluations of social and political issues: Perkins, 1985.
175 Facts, opinions, and evaluating competing evidence: Kuhn & Franklin, 2006.
History of understanding the brain: Zimmer, 2004.
The teachability of critical thinking skills: Solon, 2003; Williams et al., 2002; Moseley et al., 2004.
Philosophy for children: Topping & Tricky, 2007.
Harry Stottlemeier's Discovery: Lipman, 1970/1982.
176 Knowing about knowing: Metcalfe & Shimamura, 1994.

References

Ainslie, G. (2001). *Breakdown of will.* Cambridge, UK: Cambridge University Press.

Aizcorbe, A. M., Kennickell, A. B., & Moore, K. B. (2003). Recent changes in U.S. family finances: Evidence from the 1998 and 2001 *Survey of Consumer Finances. Federal Reserve Bulletin, 89*(1), 1–32.

Alicke, M. D., Klotz, M. L., Breitenbecher, D. L., Yurak, T. J., & Vredenburg, D. S. (1995). Personal contact, individuation, and the better-than-average effect. *Journal of Personality and Social Psychology, 68*(5), 804–25.

Allais, M. (1953). Le comportment de l'homme rationnel devant le risque: Critique des postulats et axiomes de l'école americaine. *Econometrica, 21,* 503–46.

Allman, J. (1999). *Evolving brains.* New York: Scientific American Library. Distributed by W. H. Freeman.

Allman, J., Hakeem, A., & Watson, K. (2002). Two phylogenetic specializations in the human brain. *Neuroscientist, 8*(4), 335–46.

Alloy, L. B., & Abramson, L. Y. (1979). Judgment of contingency in depressed and nondepressed students: Sadder but wiser? *Journal of Experimental Psychology, 108*(4), 441–85.

Anderson, J. R. (1990). *The adaptive character of thought.* Hillsdale, NJ: Erlbaum Associates.

Ariely, D., Loewenstein, G., & Prelec, D. (2006). Tom Sawyer and the construction of value. *Journal of Economic Behavior and Organization, 60,* 1–10.

Arkes, H. R. (1991). Costs and benefits of judgment errors: Implications for debiasing. *Psychological Bulletin, 110*(3), 486–98.

Bargh, J. A., Chen, M., & Burrows, L. (1996). Automaticity of social behavior: Direct effects of trait construct and stereotype activation on action. *Journal of Personality and Social Psychology, 71*(2), 230–44.

Bateson, M., Nettle, D., & Roberts, G. (2006). Cues of being watched enhance cooperation in a real-world setting. *Biology Letters, 2*(3), 412–14.

Bechara, A., Tranel, D., & Damasio, H. (2000). Characterization of the decision-making deficit of patients with ventromedial prefrontal cortex lesions. *Brain, 123*(11), 2189–202.

Berscheid, E., Graziano, W., Monson, T., & Dermer, M. (1976). Outcome dependency: Attention, attribution, and attraction. *Journal of Personality and Social Psychology, 34*(5), 978–89.

Blanton, H., & Gerrard, M. (1997). Effect of sexual motivation on men's risk perception for sexually transmitted disease: There must be 50 ways to justify a lover. *Health Psychology, 16*(4), 374–79.

Brickman, P., & Campbell, D. T. (1971). Hedonic relativism and planning the good society. In M. Appley (Ed.), *Adaptation-level theory* (pp. 287–305). New York: Academic Press.

Brickman, P., Coates, D., & Janoff-Bulman, R. (1978). Lottery winners and accident victims: Is happiness relative? *Journal of Personality and Social Psychology, 36*(8), 917–27.

Brown, J. C., & Loglan Institute. (1975). *Loglan I: A logical language* (3rd ed.). Gainsville, FL: Loglan Institute.

Brown, J. D. (1986). Evaluations of self and others: Self-enhancement biases in social judgments. *Social Cognition, 4*(4), 353–76.

Brown, M., & Seaton, S. (1984). *Christmas truce.* New York: Hippocrene Books.

Butler, D., Ray, A., & Gregory, L. (1995). *America's dumbest criminals.* Nashville, TN: Rutledge Hill Press.

Chater, N., Tenenbaum, J. B., & Yuille, A. (2006). Probabilistic models of cognition: Conceptual foundations. *Trends in Cognitive Science, 10*(7), 287–91.

Cheever, J. (1990, August 13). Journals. *The New Yorker.*

Chimpanzee Sequencing and Analysis Consortium. (2005). Initial sequence of the chimpanzee genome and comparison with the human genome. *Nature, 437*(7055), 69–87.

Chomsky, N. A. (1995). *The minimalist program.* Cambridge, MA: MIT Press.

Chomsky, N. A. (2000). *New horizons in the study of language and mind.* Cambridge, UK: Cambridge University Press.

Cialdini, R. B. (1993). *Influence: The psychology of persuasion*. New York: Morrow.

Clark, A. (1987). The kludge in the machine. *Mind and Language, 2,* 277–300.

Cushing, S. (1994). *Fatal words: Communication clashes and aircraft crashes.* Chicago: University of Chicago Press.

Daly, M., & Wilson, M. (1988). *Homicide.* New York: de Gruyter.

Darley, J. M., & Gross, P. H. (1983). A hypothesis-confirming bias in labeling effects. *Journal of Personality and Social Psychology, 44*(1), 20–33.

Dawkins, R. (1976). *The selfish gene.* New York: Oxford University Press.

Dawkins, R. (1982). *The extended phenotype: The gene as the unit of selection.* Oxford, UK, and San Francisco, CA: W. H. Freeman.

Dawkins, R. (1996). *Climbing Mount Improbable.* New York: Norton.

Debiec, J., Doyere, V., Nader, K., & Ledoux, J. E. (2006). Directly reactivated, but not indirectly reactivated, memories undergo reconsolidation in the amygdala. *Proceedings of the National Academy of Science USA, 103*(9), 3428–33.

Demonet, J. F., Thierry, G., & Cardebat, D. (2005). Renewal of the neurophysiology of language: Functional neuroimaging. *Physiological Reviews, 85*(1), 49–95.

Dennett, D. C. (1995). *Darwin's dangerous idea: Evolution and the meanings of life.* New York: Simon & Schuster.

Dijksterhuis, A., & Nordgren, L. F. (2006). A theory of unconscious thought. *Perspectives on Psychological Science, 1*(2), 95–109.

Dijksterhuis, A., & van Knippenberg, A. (1998). The relation between perception and behavior, or how to win a game of trivial pursuit. *Journal of Personality and Social Psychology, 74*(4), 865–77.

Dion, K. K. (1972). Physical attractiveness and evaluation of children's transgressions. *Journal of Personality and Social Psychology, 24*(2), 207–13.

Ditto, P. H., Pizarro, D. A., Epstein, E. B., Jacobson, J. A., & MacDonald, T. K. (2006). Visceral influences on risk-taking behavior. *Journal of Behavioral Decision Making, 19*(2), 99–113.

Dunning, D., Meyerowitz, J. A., & Holzberg, A. D. (1989). Ambiguity and self-evaluation: The role of idiosyncratic trait definitions in self-serving assessments of ability. *Journal of Personality and Social Psychology, 57*(6), 1082–90.

Easterlin, R. A. (1995). Will raising the incomes of all increase the happiness of all? *Journal of Economic Behavior and Organization, 27*(1), 35–47.

Epley, N., & Gilovich, T. (2006). The anchoring-and-adjustment heuristic: Why the adjustments are insufficient. *Psychological Science,* 17(4), 311–18.

Epley, N., Keysar, B., Van Boven, L., & Gilovich, T. (2004). Perspective taking as egocentric anchoring and adjustment. *Journal of Personality and Social Psychology,* 87(3), 327–39.

Epstein, S. (1994). Integration of the cognitive and the psychodynamic unconscious. *American Psychologist,* 49(8), 709–24.

Epstein, S., Lipson, A., Holstein, C., & Huh, E. (1992). Irrational reactions to negative outcomes: Evidence for two conceptual systems. *Journal of Personality and Social Psychology,* 62(2), 328–39.

Etcoff, N. L. (1999). *Survival of the prettiest: The science of beauty.* New York: Doubleday.

Fazio, R. H. (1986). How do attitudes guide behavior? In R. M. Sorrentino & E. T. Higgins (Eds.), *Handbook of motivation and cognition: Foundations of social behavior* (pp. 1, 204–33). New York: Guilford Press.

Fedde, M. R., Orr, J. A., Shams, H., & Scheid, P. (1989). Cardiopulmonary function in exercising bar-headed geese during normoxia and hypoxia. *Respiratory Physiology and Neurobiology,* 77(2), 239–52.

Ferreira, F., Bailey, K.G.D., & Ferraro, V. (2002). Good-enough representations in language comprehension. *Current Directions in Psychological Science,* 11(1), 11–15.

Ferreira, M. B., Garcia-Marques, L., Sherman, S. J., & Sherman, J. W. (2006). Automatic and controlled components of judgment and decision making. *Journal of Personality and Social Psychology,* 91(5), 797–813.

Festinger, L., & Carlsmith, J. M. (1959). Cognitive consequences of forced compliance. *Journal of Abnormal Psychology,* 58(2), 203–10.

Finlay, B. L., & Darlington, R. B. (1995). Linked regularities in the development and evolution of mammalian brains. *Science,* 268(5217), 1578–84.

Fishbach, A., Shah, J. Y., & Kruglanski, A. W. (2004). Emotional transfer in goal systems. *Journal of Experimental Social Psychology,* 40, 723–38.

Fitch, W. T. (2005). The evolution of music in comparative perspective. *Annals of the New York Academy of Sciences,* 1060(1), 29–49.

Fogg, B. J., Soohoo, C., Danielson, D., Marable, L., Stanford, J., & Tauber, E. (2002). How do people evaluate a web site's credibility?: Results from a large study. From http://www.consumerwebwatch.org/news/report3_credibilityresearch/stanfordPTL.pdf.

Forer, B. R. (1949). The fallacy of personal validation: A classroom demonstration of gullibility. *Journal of Abnormal and Social Psychology,* 44, 118–23.

Förster, J., & Strack, F. (1998). Motor actions in retrieval of valenced information: II. Boundary conditions for motor congruence effects. *Perceptual and Motor Skills, 86*(3, Pt. 2), 1423–6.

Frank, R. H. (2001). *Luxury fever: Why money fails to satisfy in an era of excess.* New York: Simon & Schuster.

Gailliot, M. T., Baumeister, R. F., DeWall, C. N., Maner, J. K., Plant, E. A., Tice, D. M., Brewer, L. E., & Schmeichel, B. J. (2007). Self-control relies on glucose as a limited energy source: Willpower is more than a metaphor. *Journal of Personality and Social Psychology, 92*(2), 325–36.

Galinsky, A. D., & Moskowitz, G. B. (2000). Counterfactuals as behavioral primes: Priming the simulation heuristic and consideration of alternatives. *Journal of Experimental Social Psychology, 36*(4), 384–409.

Galvan, A., Hare, T. A., Parra, C. E., Penn, J., Voss, H., Glover, G., & Casey, B. J. (2006). Earlier development of the accumbens relative to orbitofrontal cortex might underlie risk-taking behavior in adolescents. *Journal of Neuroscience, 26*(25), 6885–92.

Gebhart, A. L., Petersen, S. E., & Thach, W. T. (2002). Role of the posterolateral cerebellum in language. *Annals of the New York Academy of Science, 978,* 318–33.

Gelman, S. A., & Bloom, P. (2007). Developmental changes in the understanding of generics. *Cognition, 105*(1), 166–83.

Gilbert, D. T., Krull, D. S., & Malone, P. S. (1990). *Journal of Personality and Social Psychology, 59*(4), 601–13.

Gilbert, D. T., Tafarodi, R. W., & Malone, P. S. (1993). You can't not believe everything you read. *Journal of Personality and Social Psychology, 65*(2), 221–33.

Godden, D. R., & Baddeley, A. D. (1975). Context-dependent memory in two natural environments: On land and underwater. *British Journal of Psychology, 66*(3), 325–31.

Goel, V. (2003). Evidence for dual neural pathways for syllogistic reasoning. *Psychologia, 32,* 301–9.

Goel, V., & Dolan, R. J. (2003). Explaining modulation of reasoning by belief. *Cognition, 87*(1), B11–22.

Goldner, E. M., Hsu, L., Waraich, P., & Somers, J. M. (2002). Prevalence and incidence studies of schizophrenic disorders: A systematic review of the literature. *Canadian Journal of Psychiatry, 47*(9), 833–43.

Goldstein, L., Pouplier, M., Chen, L., Saltzman, E., & Byrd, D. (2007). Dynamic action units slip in speech production errors. *Cognition, 103*(3), 396–412.

Gollwitzer, P. M., & Sheeran, P. (2006). Implementation intentions and goal achievement: A meta-analysis of effects and processes. *Advances in Experimental Social Psychology, 38*, 69–119.

Graham, L., & Metaxas, P. T. (2003). "Of course it's true: I saw it on the Internet!" Critical thinking in the Internet era. *Communications of the ACM, 46*(5), 70–75.

Greene, J. D., Nystrom, L. E., Engell, A. D., Darley, J. M., & Cohen, J. D. (2004). The neural bases of cognitive conflict and control in moral judgment. *Neuron, 44*(2), 389–400.

Greene, J. D., Sommerville, R. B., Nystrom, L. E., Darley, J. M., & Cohen, J. D. (2001). An fMRI investigation of emotional engagement in moral judgment. *Science, 293*(5537), 2105–8.

Greenwald, A. G., McGhee, D. E., & Schwartz, J.L.K. (1998). Measuring individual differences in implicit cognition: The implicit association test. *Journal of Personality and Social Psychology, 74*(6), 1464–80.

Gregory, T. R. (2005). *The evolution of the genome.* Burlington, MA: Elsevier Academic.

Groopman, J. E. (2007). *How doctors think.* Boston: Houghton Mifflin.

Haidt, J. (2001). The emotional dog and its rational tail: A social intuitionist approach to moral judgment. *Psychological Review, 108*(4), 814–34.

Haselton, M. G., & Buss, D. M. (2000). Error management theory: A new perspective on biases in cross-sex mind reading. *Journal of Personality and Social Psychology, 78*(1), 81–91.

Hauser, M. D., Chomsky, N., & Fitch, W. T. (2002). The faculty of language: What is it, who has it, and how did it evolve? *Science, 298*(5598), 1569–79.

Herrnstein, R. J., & Prelec, D. (1992). In G. Loewenstein & J. Elster (Eds.), *A theory of addiction: Choice over time* (pp. 331–60). New York: Russell Sage.

Higgins, E. T. (2000). Making a good decision: Value from fit. *American Psychologist, 55*(11), 1217–30.

Higgins, E. T., Rholes, W. S., & Jones, C. R. (1977). Category accessibility and impression formation. *Journal of Experimental Social Psychology, 13*(2), 141–54.

Hoch, S. J. (1985). Counterfactual reasoning and accuracy in predicting personal events. *Journal of Experimental Psychology: Learning, Memory, and Cognition, 11*(4), 719–31.

Hornstein, H. A., LaKind, E., Frankel, G., & Manne, S. (1975). Effects of knowledge about remote social events on prosocial behavior, social conception, and mood. *Journal of Personality and Social Psychology, 32*(6), 1038–46.

Howard, S. (2004, November 14). Dreaming of sex costs the nation £7.8bn a year. *Sunday Times* (London).

Jacob, F. (1977). Evolution and tinkering. *Science, 196,* 1161–66.

Jacoby, L. L., Kelley, C., Brown, J., & Jasechko, J. (1989). Becoming famous overnight: Limits on the ability to avoid unconscious influences of the past. *Journal of Personality and Social Psychology, 56*(3), 326–38.

Jones-Lee, M., & Loomes, G. (2001). Private values and public policy. In E. U. Weber, J. Baron, & G. Loomes (Eds.), *Conflict and tradeoffs in decision making* (pp. 205–30). Cambridge, UK: Cambridge University Press.

Jost, J. T., & Hunyady, O. (2003). The psychology of system justification and the palliative function of ideology. *European Review of Social Psychology, 13*(1), 111–53.

Kagel, J. H., Green, L., & Caraco, T. (1986). When foragers discount the future: Constraint or adaptation? *Animal Behaviour, 34*(1), 271–83.

Kahneman, D., Krueger, A. B., Schkade, D. A., Schwarz, N., & Stone, A. A. (2004). A survey method for characterizing daily life experience: The day reconstruction method. *Science, 306*(5702), 1776–80.

Kahneman, D., Krueger, A. B., Schkade, D., Schwarz, N., & Stone, A. A. (2006). Would you be happier if you were richer?: A focusing illusion. *Science, 312*(5782), 1908–10.

Kahneman, D., & Ritov, I. (1994). Determinants of stated willingness to pay for public goods: A study in the headline method. *Journal of Risk and Uncertainty, 9*(1), 5–38.

Kassarjian, H. H., & Cohen, J. B. (1965). Cognitive dissonance and consumer behavior. *California Management Review, 8,* 55–64.

Kelly, A. V. (2001, January 19). What did Hitler do in the war, Miss? *Times Educational Supplement,* p. 12.

Kessler, R. C., Berglund, P., Demler, O., Jin, R., Merikangas, K. R., & Walters, E. E. (2005). *Lifetime prevalence and age-of-onset distributions of* DSM-IV *disorders in the* National Comorbidity Survey Replication. Chicago: American Medical Association.

Keysar, B., & Henly, A. S. (2002). Speakers' overestimation of their effectiveness. *Psychological Science, 13*(3), 207–12.

King, M. C., & Wilson, A. C. (1975). Evolution at two levels in humans and chimpanzees. *Science, 188*(4184), 107–16.

Kirscht, J. P., Haefner, D. P., Kegeles, S. S., & Rosenstock, I. M. (1966). A national study of health beliefs. *Journal of Health and Human Behavior, 7*(4), 248–54.

Klauer, K. C., Musch, J., & Naumer, B. (2000). On belief bias in syllogistic reasoning. *Psychological Review, 107*(4), 852–84.

Koehler, D. J. (1994). Hypothesis generation and confidence in judgment. *Journal of Experimental Psychology: Learning, Memory, and Cognition, 20*(2), 461–69.

Koriat, A., Lichtenstein, S., & Fischhoff, B. (1980). Reasons for overconfidence. *Journal of Experimental Psychology: Human Learning and Memory, 6*, 107–18.

Kray, L. J., Galinsky, A. D., & Wong, E. M. (2006). Thinking within the box: The relational processing style elicited by counterfactual mind-sets. *Journal of Personality and Social Psychology, 91*, 33–48.

Kuhn, D. (2005). *Education for thinking.* Cambridge, MA: Harvard University Press.

Kuhn, D., & Franklin, S. (2006). The second decade: What develops (and how). In W. Damon & R. Lerner (Series Eds.), D. Kuhn & R. Siegler (Vol. Eds.), *Handbook of child psychology* (pp. 953–94). New York: Wiley.

Kunda, Z. (1990). The case for motivated reasoning. *Psychological Bulletin, 108*(3), 480–98.

Larrick, R. P. (2004). Debiasing. In D. Koehler & N. Harvey (Eds.), *The Blackwell Handbook of Judgment and Decision Making* (pp. 316–37). Malden, MA: Blackwell.

Layard, P.R.G. (2005). *Happiness: Lessons from a new science.* New York: Penguin.

Leary, M. R., & Forsyth, D. R. (1987). Attributions of responsibility for collective endeavors. In C. Hendrick (Ed.), *Group processes: Review of personality and social psychology,* Vol. 8 (pp. 167–88). Thousand Oaks, CA: Sage.

Ledoux, J. E. (1996). *The emotional brain: The mysterious underpinnings of emotional life.* New York: Simon & Schuster.

Lerner, M. J. (1980). *The belief in a just world: A fundamental delusion.* New York: Plenum Press.

Leslie, S.-J. (2007). Generics and the structure of the mind. *Philosophical Perspectives, 21*(1), 378–403.

Liberman, N., Sagristano, M. D., & Trope, Y. (2002). The effect of temporal distance on level of mental construal. *Journal of Experimental Social Psychology, 38*(6), 523–34.

Lieberman, P. (1984). *The biology and evolution of language.* Cambridge, MA: Harvard University Press.

Linden, D. J. (2007). *The accidental mind.* Cambridge, MA: Belknap Press of Harvard University Press.

Linley, P. A., & Joseph, S. (2004). Positive change following trauma and adversity: A review. *Journal of Traumatic Stress, 17*(1), 11–21.

Lipman, M. (1970/1982). *Harry Stottlemeier's discovery.* Montclair, NJ: Institute for the Advancement of Philosophy for Children (IAPC).

Loftus, E. F. (2003). Make-believe memories. *American Psychologist, 58*(11), 867–73.

Lord, C. G., Ross, L., & Lepper, M. R. (1979). Biased assimilation and attitude polarization: The effects of prior theories on subsequently considered evidence. *Journal of Personality and Social Psychology, 37*(11), 2098–109.

Luria, A. K. (1971). Towards the problem of the historical nature of psychological processes. *International Journal of Psychology, 6*(4), 259–72.

Lynch Jr., J. G., & Zauberman, G. (2006). When do you want it?: Time, decisions, and public policy. *Journal of Public Policy and Marketing, 25*(1), 67–78.

Lyubomirsky, S., Caldwell, N. D., & Nolen-Hoeksema, S. (1998). Effects of ruminative and distracting responses to depressed mood on retrieval of autobiographical memories. *Journal of Personality and Social Psychology, 75*(1), 166–77.

Macrae, C. N., Bodenhausen, G. V., Milne, A. B., & Jetten, J. (1994). Out of mind but back in sight: Stereotypes on the rebound. *Journal of Personality and Social Psychology, 67*(5), 808–17.

Marcus, G. F. (1989). The psychology of belief revision. Bachelor's thesis, Hampshire College, Amherst, MA.

Marcus, G. F. (2004). *The birth of the mind: How a tiny number of genes creates the complexities of human thought.* New York: Basic Books.

Marcus, G. F., & Wagers, M. (under review). Tree structure and the structure of sentences: A reappraisal. New York University.

Marks, I., & Nesse, R. (1997). Fear and fitness: An evolutionary analysis of anxiety disorders. In S. Baron-Cohen (Ed.), *The maladapted mind: Classic readings in evolutionary psychopathology* (pp. 57–72). Hove, UK: Psychology Press.

Markus, G. B. (1986). Stability and change in political attitudes: Observed, recalled, and "explained." *Political Behavior, 8*(1), 21–44.

McClure, S. M., Botvinick, M. M., Yeung, N., Greene, J. D., & Cohen, J. D. (in press). Conflict monitoring in cognition-emotion competition. In J. J. Gross (Ed.), *Handbook of emotion regulation.* New York: Guilford.

Mealey, L. (1995). The sociobiology of sociopathy: An integrated evolutionary model. *Behavioral and Brain Sciences, 18*(3), 523–41.

Messick, D. M., Bloom, S., Boldizar, J. P., & Samuelson, C. D. (1985). Why we are fairer than others. *Journal of Experimental Social Psychology, 21*(5), 480–500.

Meston, C. M., & Buss, D. M. (2007). Why humans have sex. *Archives of Sexual Behavior, 36*(4), 477–507.

Metcalfe, J., & Shimamura, A. P. (1994). *Metacognition: Knowing about knowing.* Cambridge, MA: MIT Press.

Metzger, M. J., Flanagin, A. J., & Zwarun, L. (2003). College student Web use, perceptions of information credibility, and verification behavior. *Computers and Education, 41*(3), 271–90.

Miller, G., & Chomsky, N. A. (1963). Finitary models of language users. In R. D. Luce, R. R. Bush, & E. Galanter (Eds.), *Handbook of mathematical psychology* (Vol. II). New York: Wiley.

Miller, G. F. (2000). Evolution of human music through sexual selection. In N. L. Wallin, B. Merker, & S. Brown (Eds.), *The origins of music* (pp. 329–60). Cambridge, MA: MIT Press.

Minino, A. M., Arias, E., Kochanek, K. D., Murphy, S. L., & Smith, B. L. (2002). Deaths: Final data for 2000. *National Vital Statistics Report, 50*(15), 1–119.

Mischel, W., Shoda, Y., & Rodriguez, M. I. (1989). Delay of gratification in children. *Science, 244*(4907), 933–38.

Montague, R. (2006). *Why choose this book?: How we make decisions.* New York: Dutton.

Montalbetti, M. M. (1984). After binding: On the interpretation of pronouns. Doctoral dissertation, MIT, Cambridge, MA.

Moseley, D., Baumfield, V., Higgins, S., Lin, M., Miller, J., Newton, D., Robson, S., Elliot, J., & Gregson, M. (2004). Thinking skill frameworks for post-16 learners: An evaluation. Newcastle upon Tyne, UK: Research Centre, School of Education.

Mussweiler, T., Strack, F., & Pfeiffer, T. (2000). Overcoming the inevitable anchoring effect: Considering the opposite compensates for selective accessibility. *Personality and Social Psychology Bulletin, 26*(9), 1142.

Nesse, R. (1997). An evolutionary perspective on panic disorder and agoraphobia. In S. Baron-Cohen (Ed.), *The maladapted mind: Classic read-*

ings in evolutionary psychopathology (pp. 72–84). Hove, UK: Psychology Press.

Nesse, R. M., & Williams, G. C. (1994). *Why we get sick: The new science of Darwinian medicine* (1st ed.). New York: Times Books.

Nickerson, R. S. (1988). On improving thinking through instruction. *Review of Research in Education, 15,* 3–57.

Nielsen. (2006). Nielsen Media Research reports television's popularity is still growing. From http://www.nielsenmedia.com/nc/portal/site/Public/menuitem.55dc65b4a7d5adff3f65936147a062a0/?vgnextoid= 4156527aacccd010VgnVCM100000ac0a260aRCRD.

Nisbett, R. E., Krantz, D. H., Jepson, C., & Kunda, Z. (1983). The use of statistical heuristics in everyday inductive reasoning. *Psychological Review, 90,* 339–63.

Noice, H., & Noice, T. (2006). What studies of actors and acting can tell us about memory and cognitive functioning. *Current Directions in Psychological Science, 15*(1), 14–18.

Nuttin, J. M. (1987). Affective consequences of mere ownership: The name letter effect in twelve European languages. *European Journal of Social Psychology, 17*(4), 381–402.

Oakhill, J., Johnson-Laird, P. N., & Garnham, A. (1989). Believability and syllogistic reasoning. *Cognition, 31*(2), 117–40.

Pacini, R., Muir, F., & Epstein, S. (1998). Depressive realism from the perspective of cognitive-experiential self-theory. *Journal of Personality and Social Psychology, 74*(4), 1056–68.

Pandelaere, M., & Dewitte, S. (2006). Is this a question? Not for long: The statement bias. *Journal of Experimental Social Psychology, 42*(4), 525–31.

Parker, A. (2006). Evolution as a constraint on theories of syntax: The case against minimalism. Doctoral dissertation, University of Edinburgh, Edinburgh, UK.

Perkins, D. N. (1985). Postprimary education has little impact on informal reasoning. *Journal of Educational Psychology, 77*(5), 562–71.

Pew Research Center. (2007). Republicans lag in engagement and enthusiasm for candidates. From http://people-press.org/reports/pdf/307.pdf.

Pinker, S., & Jackendoff, R. (2005). The faculty of language: What's special about it? *Cognition, 95*(2), 201–36.

Plomin, R. (1997). *Behavioral genetics* (3rd ed.). New York: W. H. Freeman.

Plomin, R., DeFries, J. C., McClearn, G. E., & McGuffin, P. (2001). *Behavior genetics.* New York: Worth.

Polimeni, J., & Reiss, J. P. (2002). How shamanism and group selection may reveal the origins of schizophrenia. *Medical Hypotheses, 58*(3), 244–48.

Posner, M. I., & Keele, S. W. (1968). On the genesis of abstract ideas. *Journal of Experimental Psychology, 77*(3), 353–63.

Prasada, S. (2000). Acquiring generic knowledge. *Trends in Cognitive Sciences, 4,* 66–72.

Premack, D. (2004). Psychology: Is language the key to human intelligence? *Science, 303*(5656), 318–20.

Price, J., Sloman, L., Russell Gardner, J., Gilbert, P., & Rohde, P. (1997). The social competition hypothesis of depression. In S. Baron-Cohen (Ed.), *The maladapted mind: Classic readings in evolutionary psychopathology* (pp. 241–54). Hove, UK: Psychology Press.

Pullum, G. K. (1991). *The great Eskimo vocabulary hoax and other irreverent essays on the study of language.* Chicago: University of Chicago Press.

Quattrone, G. A., & Tversky, A. (1988). Contrasting rational and psychological analyses of political choice. *American Political Science Review, 82,* 719–36.

Rachlin, H. (2000). *The science of self-control.* Cambridge, MA: Harvard University Press.

Read, D., & van Leeuwen, B. (1998). Predicting hunger: The effects of appetite and delay on choice. *Organizational Behavior and Human Decision Processes, 76*(2), 189–205.

Reder, L. M., & Kusbit, G. W. (1991). Locus of the Moses illusion: Imperfect encoding, retrieval, or match. *Journal of Memory and Language, 30,* 385–406.

Robinson, T. N., Borzekowski, D.L.G., Matheson, D. M., & Kraemer, H. C. (2007). Effects of fast-food branding on young children's taste preferences. *Archives of Pediatrics and Adolescent Medicine, 161*(8), 792.

Rosa-Molinar, E., Krumlauf, R. K., & Pritz, M. B. (2005). Hindbrain development and evolution: Past, present, and future. *Brain, Behavior, and Evolution, 66*(4), 219–21.

Ross, M., & Sicoly, F. (1979). Egocentric biases in availability and attribution. *Journal of Personality and Social Psychology, 37*(3), 322–36.

Russell, B. (1918/1985). *The philosophy of logical atomism.* Lasalle, IL: Open Court.

Russo, J. E., & Schoemaker, P.J.H. (1989). *Decision traps: Ten barriers to brilliant decision-making and how to overcome them* (1st ed.). New York: Doubleday/Currency.

Schacter, D. L. (2001). *The seven sins of memory: How the mind forgets and remembers.* Boston: Houghton Mifflin.

Schacter, D. L., & Addis, D. R. (2007). Constructive memory: The ghosts of past and future. *Nature, 445*(7123), 27.

Schelling, T. C. (1984). *Choice and consequence.* Cambridge, MA: Harvard University Press.

Schooler, J. W., Reichle, E. D., & Halpern, D. V. (2004). Zoning out while reading: Evidence for dissociations between experience and meta-consciousness. In *Thinking and seeing: Visual metacognition in adults and children* (pp. 203–206). Cambridge, MA: MIT Press.

Schwartz, B., & Schwartz, B. (2004). *Paradox of choice: Why more is less.* New York: HarperCollins.

Schwarz, N., Strack, F., & Mai, H. P. (1991). Assimilation and contrast effects in part-whole question sequences: A conversational logic analysis. *Public Opinion Quarterly, 55*(1), 3–23.

Sherman, J. W., Macrae, C. N., & Bodenhausen, G. V. (2000). Attention and stereotyping: Cognitive constraints on the construction of meaningful social impressions. *European Review of Social Psychology, 11*, 145–75.

Shiv, B., & Fedorikhin, A. (1999). Heart and mind in conflict: The interplay of affect and cognition in consumer decision making. *Journal of Consumer Research, 26*(3), 278.

Simon, L., Greenberg, J., Harmon-Jones, E., Solomon, S., Pyszczynski, T., Arndt, J., & Abend, T. (1997). Terror management and cognitive-experiential self-theory: Evidence that terror management occurs in the experiential system. *Personality and Social Psychology, 72*(5), 1132–46.

Simons, D. J., & Levin, D. T. (1998). Failure to detect changes to people during a real-world interaction. *Psychonomic Bulletin and Review, 5*(4), 644–49.

Smith, D. M., Schwarz, N., Roberts, T. R., & Ubel, P. A. (2006). Why are you f=calling me?: How study introductions change response patterns. *Quality of Life Research, 15*(4), 621–30.

Smolin, L. (2006). *The trouble with physics: The rise of string theory, the fall of a science, and what comes next.* Boston: Houghton Mifflin.

Solomon, S., Greenberg, J., & Pyszczynski, T. (2004). The cultural animal: Twenty years of terror-management theory and research. *Handbook of Experimental Existential Psychology, 13*–34.

Solon, T. (2003). Teaching critical thinking!: The more, the better. *The Community College Experience, 9*(2), 25–38.

Stanovich, K. E. (2003). The fundamental computational biases of human cognition: Heuristics that (sometimes) impair decision making and problem solving. In J. E. Davidson & R. J. Sternberg (Eds.), *The psychology of problem solving* (pp. 291–342). New York: Cambridge University Press.

Steel, P. (2007). The nature of procrastination: A meta-analytic and theoretical review of quintessential self-regulatory failure. *Psychological Bulletin, 133*(1), 65–94.

Steele, C. M., & Aronson, J. (1995). Stereotype threat and the intellectual test performance of African Americans. *Journal of Personality and Social Psychology 69*(5), 797–811.

Stich, S. (in press). Nicod lectures on morality. Cambridge, MA: MIT Press. Videos available at semioweb.msh-paris.fr/AR/974/liste_conf.asp.

Strack, F., Martin, L. L., & Schwarz, N. (1988). Priming and communication: Social determinants of information use in judgments of life satisfaction. *European Journal of Social Psychology, 18*(5), 429–42.

Strack, F., Martin, L. L., & Stepper, S. (1988). Inhibiting and facilitating conditions of the human smile: A nonobtrusive test of the facial feedback hypothesis. *Journal of Personality and Social Psychology, 54*(5), 768–77.

Svenson, O. (1981). Are we all less risky and more skillful than our fellow drivers? *Acta Psychologica, 47*(2), 143–48.

Takahashi, T. (2005). The evolutionary origins of vertebrate midbrain and MHB: Insights from mouse, amphioxus and ascidian Dmbx homeobox genes. *Brain Research Bulletin, 66*(4–6), 510–17.

Talarico, J. M., & Rubin, D. C. (2003). Confidence, not consistency, characterizes flash-bulb memories. *Psychological Science, 14*(5), 455–61.

Tetlock, P. E. (1985). Accountability: A social check on the fundamental attribution error. *Social Psychology Quarterly, 48*(3), 227–36.

Thaler, R. H. (1999). Mental accounting matters. *Journal of Behavioral Decision Making, 12*(3), 183–206.

Thompson, C. (2007). Halo 3: How Microsoft labs invented a new science of play. *Wired, 15*, 140–47.

Thomson, J. J. (1985). The trolley problem. *Yale Law Journal, 94*(6), 1395–415.

Todorov, A., Mandisodza, A. N., Goren, A., & Hall, C. C. (2005). Inferences of competence from faces predict election outcomes. *Science, 308*(5728), 1623–6.

Tooby, J., & Cosmides, L. (1995). Mapping the evolved functional organization of mind and brain. In M. S. Gazzaniga (Ed.), *The cognitive neurosciences* (pp. 1185–97). Cambridge, MA: MIT Press.

Topping, K. J., & Trickey, S. (2007). Collaborative philosophical enquiry for school children: Cognitive effects at 1012 years. *British Journal of Educational Psychology, 77*(2), 271–88.

Trehub, S. (2003). Musical predispositions in infancy: An update. In I. Peretz & R. J. Zattore (Eds.), *The cognitive neuroscience of music* (pp. 3–20). New York: Oxford University Press.

Trivers, R. (1972). *Parental investment and sexual selection.* Oxford, UK: Oxford University Press.

Tuchman, B. (1984). *The march of folly: From Troy to Vietnam* (1st ed.). New York: Knopf.

Tulving, E., & Craik, F.I.M. (2000). *The Oxford handbook of memory.* New York: Oxford University Press.

Tversky, A., & Kahneman, D. (1974). Judgment under uncertainty: Heuristics and biases. *Science, 185*(4157), 1124–31.

Tversky, A., & Kahneman, D. (1981). The framing of decisions and the psychology of choice. *Science, 211*(4481), 453–8.

Tyre, P. (2004, June 7). Clean freaks. *Newsweek.*

U.S. Department of Labor Statistics. (2007, June 28). American time use survey summary. From http://www.bls.gov/news.release/atus.nro.htm.

Wansink, B., Kent, R. J., & Hoch, S. J. (1998). An anchoring and adjustment model of purchase quantity decisions. *Journal of Marketing Research, 35*(1), 71–81.

Wason, P. C. (1960). On the failure to eliminate hypotheses in a conceptual task. *Quarterly Journal of Experimental Psychology, 12*, 129–40.

Watkins, P. C., Vache, K., Verney, S. P., Muller, S., & Mathews, A. (1996). Unconscious mood-congruent memory bias in depression. *Journal of Abnormal Psychology, 105*(1), 34–41.

Wegner, D. M. (1994). Ironic processes of mental control. *Psychological Review, 101*(1), 34–52.

Weiner, J. (1994). *The beak of the finch: A story of evolution in our time* (1st Vintage Books ed.). New York: Vintage Books.

Wesson, R. G. (1991). *Beyond natural selection.* Cambridge, MA: MIT Press.

Williams, W. M., Blythe, T., White, N., Li, J., Gardner, H., & Sternberg, R. J. (2002). Practical intelligence for school: Developing metacognitive sources of achievement in adolescence. *Developmental Review, 22*(2), 162–210.

Wilson, T. D., & Brekke, N. (1994). Mental contamination and mental correction: Unwanted influences on judgments and evaluations. *Psychological Bulletin, 116*(1), 117–42.

Winkielman, P., & C. Berridge, K. (2004). Unconscious emotion. *Current Directions in Psychological Science, 13*(3), 120–3.

Zajonc, R. B. (1968). Attitudinal effects of mere exposure. *Journal of Personality and Social Psychology, 9*(2, Pt. 2), 1–27.

Zimmer, C. (2004). *Soul made flesh: The discovery of the brain — and how it changed the world.* New York: Free Press.

Index

Westfield Memorial Library
Westfield, New Jersey